UNDERSTANDING STRUCTURES
An Introduction to Structural Analysis

UNDERSTANDING STRUCTURES
An Introduction to Structural Analysis

Mete A. Sozen
Toshikatsu Ichinose

CRC Press is an imprint of the
Taylor & Francis Group, an **informa** business

CRC Press
Taylor & Francis Group
6000 Broken Sound Parkway NW, Suite 300
Boca Raton, FL 33487-2742

© 2009 by Taylor & Francis Group, LLC
CRC Press is an imprint of Taylor & Francis Group, an Informa business

No claim to original U.S. Government works
Printed in the United States of America on acid-free paper
10 9 8 7 6 5 4 3 2 1

International Standard Book Number-13: 978-1-4200-6861-0 (Hardcover)

This book contains information obtained from authentic and highly regarded sources. Reasonable efforts have been made to publish reliable data and information, but the author and publisher cannot assume responsibility for the validity of all materials or the consequences of their use. The authors and publishers have attempted to trace the copyright holders of all material reproduced in this publication and apologize to copyright holders if permission to publish in this form has not been obtained. If any copyright material has not been acknowledged please write and let us know so we may rectify in any future reprint.

Except as permitted under U.S. Copyright Law, no part of this book may be reprinted, reproduced, transmitted, or utilized in any form by any electronic, mechanical, or other means, now known or hereafter invented, including photocopying, microfilming, and recording, or in any information storage or retrieval system, without written permission from the publishers.

For permission to photocopy or use material electronically from this work, please access www.copyright.com (http://www.copyright.com/) or contact the Copyright Clearance Center, Inc. (CCC), 222 Rosewood Drive, Danvers, MA 01923, 978-750-8400. CCC is a not-for-profit organization that provides licenses and registration for a variety of users. For organizations that have been granted a photocopy license by the CCC, a separate system of payment has been arranged.

Trademark Notice: Product or corporate names may be trademarks or registered trademarks, and are used only for identification and explanation without intent to infringe.

Library of Congress Cataloging-in-Publication Data

Sozen, Mete Avni, 1930-
  Understanding structures : an introduction to structural analysis / Mete A. Sozen, Toshikatsu Ichinose.
    p. cm.
  Includes bibliographical references and index.
  ISBN 978-1-4200-6861-0 (hardback : alk. paper)
  1. Structural analysis (Engineering)--Data processing. 2. GOYA (Electronic resource) I. Ichinose, Toshikatsu. II. Title.

TA647.S68 2008
624.1'71--dc22   2008008861

Visit the Taylor & Francis Web site at
http://www.taylorandfrancis.com

and the CRC Press Web site at
http://www.crcpress.com

Contents

Chapter 1 Trusses 1
1.1   What Is a Truss? 1
1.2   Bar in Tension or Compression 3
1.3   Symmetrical Truss with Two Elements of Equal Size at 90° 16
1.4   Symmetrical Truss with Two Elements at Various Orientations 25
1.5   Unsymmetrical Truss with Two Elements 31
1.6   Truss with Three Members 39
1.7   Hands-on Approach to Truss Design 51
1.8   Analysis of Statically Determinate Trusses 58
1.9   Stable Trusses 68
1.10 Building a Truss 72
1.11 Problems 73

Chapter 2 Moments and Deflections in Cantilever Beams 79
2.1   What Is a Bending Moment? 79
2.2   What Is a Shear Force? 89
2.3   What Is a Distributed Load? 106
2.4   What Is a Couple? 118
2.5   The Effects of Moment on Stresses in a Beam 123
2.6   The Effects of Moment on Strains in a Beam 130
2.7   Deformation of a Beam—Spring Model 135
2.8   Deformation of a Beam—Continuously Deformable Model 143
2.9   Problems 156

Chapter 3 Moments and Deflections in Simply Supported Beams 161
3.1   The Effect of a Concentrated Load on Shear, Moment, and Deflection 161
3.2   The Effect of Several Concentrated Loads on Shear, Moment,
      and Deflection 171
3.3   Similarities between Beam and Truss Response 181
3.4   Construction and Test of a Timber Beam 188
3.5   Problems 194

Chapter 4 Bending and Shear Stresses 199
4.1   First Moment 199
4.2   Second Moment and Section Modulus 206
4.3   Construction and Test of a Styrofoam Beam 216
4.4   Shear Stress 218
4.5   Problems 225

Chapter 5 Frames 227
5.1 Introductory Concepts.. 227
5.2 A Simple Bent 236
5.3 A Portal Frame 251
5.4 Statically Indeterminate Frame.. 261
5.5 Multistory Frame 271
5.6 Three-Hinged Frame.. 279
5.7 Problems... 286

Chapter 6 Buckling 291
6.1 Simple Models.. 291
6.2 Continuously Deformable Model... 302
6.3 Problems... 309

Answers to Problems 311

List of Symbols 341
Users Manual for GOYA 341

Index 347

Preface

This book is committed to the development of an intuitive understanding of structural response. The material in the book brings together the art and science of structures in the environment of a computer game. The software GOYA, specially developed for this purpose, is an integral part of the book. Access one of the following websites to download GOYA.

http://kitten.ace.nitech.ac.jp/ichilab/mech/en/
ftp://ftp.ecn.purdue.edu/sozen/GOYA

Structural behavior is best learned by repeated observation and, sometimes, by intensely bad experience. The old masters, who built elegant and efficient structures long before the first glimmerings of the theory of elasticity, demonstrated that quite well. Modern life does not encourage long apprenticeships. The art of analysis for design needs to be learned in a short period of time. To reach maturity in understanding structures requires the development of a sense of proportion and an innate knowledge of how structural elements deform. GOYA will supply instant answers to the questions that the reader may ask: Are the members strong enough to carry the applied load? What if the cross-sectional area of the member is changed? What is the effect of the change on the deformed shape? Is the member still strong enough to resist the load? What if the loading differs from the one first assumed? How is the deflected shape affected?

The reader can develop judgment by seeing graphic and quantitative answers to such questions very quickly and conveniently. It is not as good as experience and certainly not as memorable as bad experience, but the answers are inexpensive and immediately available. Simple physical experiments are also prescribed to reinforce intuitive understanding of structural response obtained from exercises with GOYA.

The book takes the reader from elementary concepts of forces in trusses to bending of beams and response of multistory, multibay frames in a carefully designed cadence. The text, the figures, and the relevant portions of GOYA create an interactive learning environment for understanding how trusses, beams, and frames work. We hope that it will entertain as well as edify the reader.

We wrote the book to be understood not only by structural engineers but also by architectural engineers and architects. In fact, it is our firm belief that anyone with a threshold understanding of physics should be able to learn structures using this book, preferably with, but also without, an instructor.

We are indebted to Dr. Yo Hibino for his contributions to GOYA throughout its development, to Tomoharu Hayase for his work during the initiation of GOYA, and to Naoya Wakita, Takashi Tozawa, and Ken Matsui who helped in various stages of development of the program. Naoya Wakita is also thanked for drawing cartoons. Dr. Santiago Pujol and Jeff Rautenberg are thanked for their critical reading of the manuscript.

Simple directions for using GOYA are included in a brief manual in pages 341–346. GOYA will operate on any platform that accommodates JAVA. JAVA can be downloaded from Java Technology (Sun Systems) on the Web.

Mete A. Sozen
Purdue University
West Lafayette, Indiana

T. Ichinose
Nagoya Institute of Technology
Nagoya, Japan

1 Trusses

1.1 WHAT IS A TRUSS?

Which of the three methods shown in Figure 1.1.1 would you prefer to use if you wanted to snap a chopstick? Probably, you will elect the method shown in Figure 1.1.1c. It is difficult to snap the chopstick if you pull or push it. The same rule applies to many things such as a chocolate bar, a cookie, and a plastic stick.

Assume that you need to cook a pot of stew in a campground. The structure shown in Figure 1.1.2a looks vulnerable because the horizontal bar bends and might break at the center. This type of structural system is called a *frame*. You will learn why and how the bar bends in Chapter 2. The triangular system shown in Figure 1.1.2b is stronger and is called a *truss*. The reason for the superiority of the truss over the frame in this application can be understood as follows. The force, P, corresponding to the pull of gravity on the pot may be resolved into two components $P/\sqrt{2}$ as shown in Figure 1.1.3a. Each of these components is collinear with an element of the truss and causes the reaction from the ground shown in Figure 1.1.3b. As a result, each element of the truss is pushed as illustrated in Figure 1.1.3c or Figure 1.1.1b and not bent as illustrated in Figure 1.1.1c.

The truss is also useful for bridge structures. You may make a bridge simply by laying a log as shown in Figure 1.1.4a. Such a bridge is flexible and might break if an elephant crosses it. A truss bridge (Figure 1.1.4b), which is made forming triangles, is much stiffer and stronger. Ancient Roman engineers already knew the fact by experience.* An Italian architect of the 16th century, Palladio, designed many truss timber bridges, some of which still remain in Venice. After the invention of mass production methods for steel in 19th century England, steel trusses became very popular for bridges. The Eiffel Tower (http://www.tour-eiffel.fr), built in 1889, is a truss. The Nagoya Dome, an elegant structure, is also a truss (Figure 1.1.5). The proposed space structure, Figure 1.1.6, is also based on the truss principle.

However, any one collection of triangles does not necessarily constitute a safe truss structure. The larger the structure you want to build, the more careful the calculations you should make. In ancient or medieval times, people constructed large structures by experience. That was a difficult and dangerous undertaking. In Ancient Egypt, several pyramids collapsed during construction.† The Cathédrale Saint-Pierre de Beauvais in France experienced two major collapses during its long history of construction: the choir in 1284 and the tower in 1573. Structures accomplished on the basis of experience related to gravity load only were also in danger of collapse if subjected to tornados or earthquakes.

---

* Timoshenko, S. P. 1953. *History of Strength of Materials*. McGraw-Hill, New York.
† Mendelssohn, Kurt. 1974. *The Riddle of the Pyramids*. Praeger, New York.

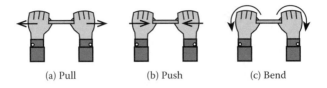

FIGURE 1.1.1 Snap a chopstick.

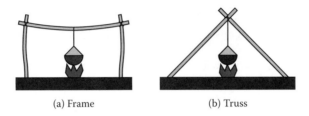

FIGURE 1.1.2 Cook a pot of stew.

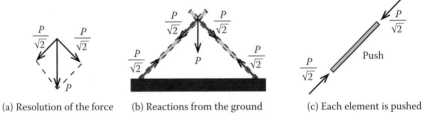

FIGURE 1.1.3 Equilibrium.

FIGURE 1.1.4 Timber bridges.

FIGURE 1.1.5 Truss in Nagoya Dome.

Trusses

FIGURE 1.1.6 Truss in a space station (http://www.nasda.go.jp/).

After structural mechanics was developed in the 19th century, the situation with respect to understanding structural behavior improved dramatically. Present-day engineers can check whether the designed structure is safe or not *before* they start construction (though the prediction is not always perfect). Structural mechanics has also reduced waste of materials and workmanship, which had been inevitable in olden times.

The investment in constructed facilities includes many components in addition to the structure. In fact, in most applications the cost of the structure is less than one-fourth of the cost of the total investment, and in industrial facilities this ratio may be even less. However, the survival of the entire investment depends on the survival of the structure. That is why the structural engineer has a critical responsibility and needs to be well informed about the behavior of the structure. The first step to develop a fundamental understanding of structural behavior is to understand mechanics, which we shall attempt to do in this book.

1.2 BAR IN TENSION OR COMPRESSION

In this section, you will study the behavior of a bar in tension and compression, before learning about a truss made up of many bars (elements). Access one of the following Web sites to start "GOYA-T."

http://kitten.ace.nitech.ac.jp/ichilab/mech/en/
ftp://ftp.ecn.purdue.edu/sozen/GOYA

You will find Figure 1.2.1, which shows all types of trusses included in this chapter.

Select the upper left figure. The left end of the bar is connected to a wall as shown in Figure 1.2.2 so that the end can rotate freely but cannot move away from the pin. Such a connection is called a *pin support*. The right end of the bar can either rotate as shown in Figure 1.2.3a or move in the horizontal direction as shown in Figure 1.2.3b but not in the vertical direction. Such a connection is called a *roller support*.*

---

* The rigorous definition of this support is shown in Figure 1.2.3c: the support has two sets of rollers so that it does not move either in the upward or downward direction. We need such a support to build a large structure. If a very long bar with pin-and-roller supports is subjected to high temperature in summer, the bar will freely elongate. If the bar has pin supports at both ends, the bar cannot elongate, and a large compressive force may destroy the bar or the supports.

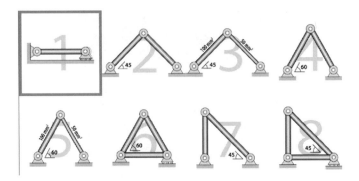

FIGURE 1.2.1 "GOYA-T" window.

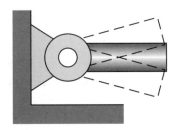

FIGURE 1.2.2 Rotation at pin support.

Drag the right end of the bar in the direction of F as shown in Figure 1.2.4a. An arrow will appear, and the end will move slightly to the right. The arrow represents the force applied to the end. The vertical component of the force, $F\sin\theta$, is supported by the floor (Figure 1.2.4a) through the rollers. In other words, the floor pushes the bar with a force of $F\sin\theta$. This force is called a *reaction*. The horizontal component, $F\cos\theta$, is transmitted through the bar and supported by the wall. In other words, the wall pulls the bar* with a force of $F\cos\theta$. This force is also a reaction. The external force F is balanced by the reactions as shown in Figure 1.2.4b. On the other hand,

---

* The force exerted by the wall on the bar acts from right to left as shown in Figure 1.1(a). The force exerted by the bar on the wall works in the opposite direction (Figure 1.1(b)). This is Newton's third law.

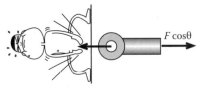

(a) Force exerted by the wall on the bar.

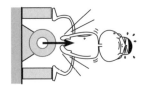

(b) Force exerted by the bar on the wall.

FIGURE 1.1 Newton's third law.

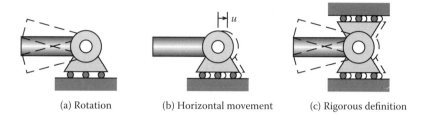

(a) Rotation (b) Horizontal movement (c) Rigorous definition

FIGURE 1.2.3 Roller supports.

the bar is pulled in tension by the force $F\cos\theta$. This action is called an *axial force* and we shall use the letter P to represent it.

There is another way to think of the meaning of axial force. Assume that we cut the bar and glue the sections as shown in Figure 1.2.5a. Then, apply the force, F, again. The glue must be strong enough to resist a force $P = F\cos\theta$ (Figure 1.2.5b,c) to satisfy the equilibrium of forces (Figure 1.2.5d). This force P is an axial force. We call this procedure an *imaginary cut*.

The roller support moves to the right because the bar elongates. Assume that the bar is made of iron. Iron is made up of atoms. Each atom deforms as shown in Figure 1.2.6. This is the reason for the lengthening of the bar.*

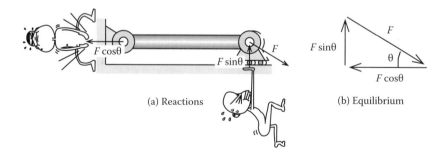

(a) Reactions (b) Equilibrium

FIGURE 1.2.4 Forces and reactions.

---

* Strictly, the atom does not deform, but the distance between the atoms changes.

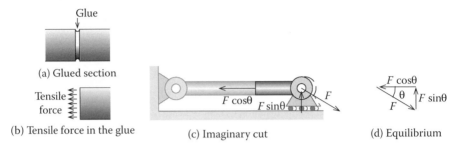

(a) Glued section

(b) Tensile force in the glue

(c) Imaginary cut

(d) Equilibrium

FIGURE 1.2.5 Equilibrium near the roller support.

To describe the deformation of any material, we will define *stress** and *strain* as follows:

$$\text{Stress: } \sigma = P/A \tag{1.2.1}$$

where P = axial force (positive in tension) and A = sectional area

$$\text{Strain: } \varepsilon = e/L \tag{1.2.2}$$

where e = deformation (positive for lengthening) and L = original length

The reason why we divide the force by the area is that the number of atoms per unit section is definite depending on the material. The stress represents the force on each atom as shown in Figure 1.2.7a. The reason for dividing the deformation by the length is that the number of atoms per unit length is definite depending on the material. The strain represents the deformation of each atom as shown in Figure 1.2.7b.

In the SI system, force is usually described using the unit N (newton). It is equal to the force that imparts an acceleration of 1 m/s² to a mass of 1 kg. A force of 1 N is approximately the gravity force generated by a mass of 0.1 kg (as an apple of 3.6 oz) because the acceleration of gravity is 9.8 m/s². The unit stress is often stated in terms of N/mm², also called a mega-Pascal and abbreviated as MPa.

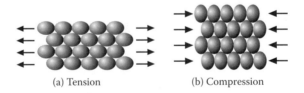

(a) Tension

(b) Compression

FIGURE 1.2.6 Deformation of atoms.

---

* We ought to call this quantity *unit stress* because the word "stress" is also used in the sense of force. However, engineers understand stress to refer to unit stress. The same explanation applies to "strain."

Trusses

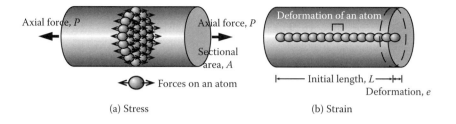

(a) Stress

(b) Strain

FIGURE 1.2.7 Definition of stress and strain.

In the imperial system of units used commonly in the United States, the unit of force is the pound, lbf, or the kip, which refers to 1000 lbf. The unit of stress is pound per square inch, psi, or ksi, which is equal to 1000 psi.

The strain, as we use it in mechanics, does not have a unit because of the way it is defined: the deformation in millimeters is divided by the length in millimeters, or a deformation stated in inches is divided by the original length in inches. If a bar of 1000 mm elongates by 0.5 mm, the strain is 0.5×10^{-3}.

Most people have an instinctive understanding of (unit) stress because of their own experiences. Consider the task of driving a large log into the ground (Figure 1.2.8a). It is difficult because the force one can muster is distributed over a large area, the cross section of the log. But if one reduces the diameter of the log (Figure 1.2.8b), which is similar to sharpening the log (Figure 1.2.8c), it becomes easier to drive it in because the force is distributed over a limited area and the stress applied on the ground is higher.

One can understand the concept of strain by thinking of the difference between pulling a short and a long rubber band. The long band stretches more even if you

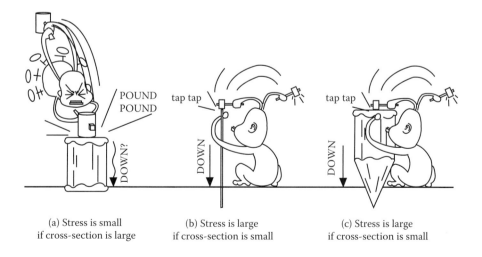

(a) Stress is small if cross-section is large

(b) Stress is large if cross-section is small

(c) Stress is large if cross-section is small

FIGURE 1.2.8 A thin stick can be easily driven in.

apply the same force (i.e., the same stress). Why? The same stress causes the same strain, ε; therefore, the deformation ($e = \varepsilon.L$), which is proportional to the original length L, is larger in the longer band. If you want to break the bands, it would take the same force, but the extension of the longer band at fracture would be again larger.

PRACTICE USING GOYA-T

Drag the right end of the bar to the left or to the right. As you drag the bar, record the extension (increase in length) and the corresponding force. Plot the force against the increase in length, e. Draw a graph of the relationship between the axial force, P, and the increase in length, e. You will obtain a plot as shown in Figure 1.2.9a. You will also notice that the bar turns blue or red if the force exceeds 100 N or −100 N. The change in color indicates that the axial force of the bar exceeds the tensile or compressive strength. In GOYA-T, the section area of the bar is assumed to be 100 mm². Look at the lower left corner showing that the tensile and compressive strengths of the material are 1 N/mm². This is why the bar appears to break at a force of 100 N or −100 N. You can change the strength of the material using the text box in the lower left corner.

The length of the bar is assumed to be 1000 mm. Because the deformation at fracture is 0.5 mm, we can determine that the strain at fracture is $0.5/1000 = 0.5 \times 10^{-3}$ for this particular material as shown in Figure 1.2.9b.

As you will learn soon, the stress–strain relationships of most materials are linear at small strains as shown in Figure 1.2.9b. We describe the relationship as follows.

$$\sigma = E.\varepsilon \tag{1.2.3}$$

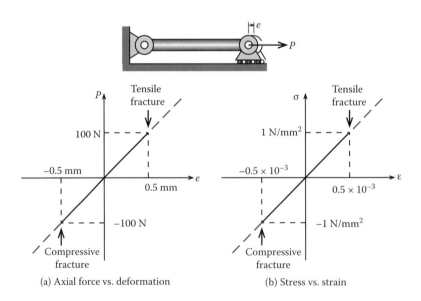

(a) Axial force vs. deformation   (b) Stress vs. strain

FIGURE 1.2.9 Horizontal force.

Trusses

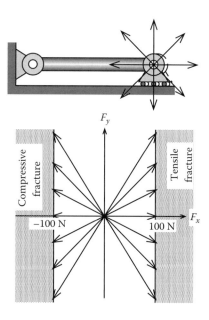

FIGURE 1.2.10 Inclined force.

The coefficient E is called Young's modulus after an English scientist, Thomas Young (1773–1859), who also contributed to archeology deciphering ancient Egyptian characters. Young's modulus is a measure of how stiff the material is, because the strain ε is inversely proportional to E. In the case of Figure 1.2.9b, $E = \sigma/\varepsilon = 2 \times 10^3$ N/mm². Note that the unit of Young's modulus is same as that of stress because strain has no unit.

Question: If the bar is made of a material with tensile and compressive strengths equal to those in Figure 1.2.9b, but if its Young's modulus is halved, how would the relationship between axial force and deformation look?

Practice using GOYA-T: Apply forces in various directions as shown in Figure 1.2.10. You will find that the bar is safe (the color does not change) if the absolute value of P_x is less than a threshold of 100 N. The vertical lines in Figure 1.2.10 show the thresholds. Change the compressive strength to 2 N/mm² while keeping the tensile strength at 1 N/mm². Show the thresholds.

We need to distinguish *stiffness* from *strength*. As J. E. Gordon described in his book,* a biscuit is stiff but weak, whereas nylon is flexible but strong. Figure 1.2.11 shows Young's modulus and tensile strengths of various materials. Tendon (bundled fibers connecting bone and muscle) is as strong as bone but much more flexible. Nylon is more flexible than timber† but ten times as strong. The strength of steel

---

* Gordon, J. E. 1978. *Structures: Or Why Things Don't Fall Down*, Da Capo Press, New York.
† Timber is stronger and stiffer when stressed in the direction of its fibers. The plot in Figure 1.2.11 shows the data for such a condition.

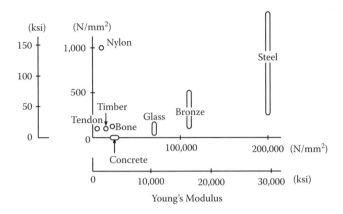

FIGURE 1.2.11 Young's modulus and tensile strength.

varies depending on its ingredients and treatment, but its Young's modulus is approximately 200 kN/mm² (29,000 ksi) whatever its strength is.

For interested readers: The stress–strain relationships of actual materials are approximately as shown in Figure 1.2.12: Most are linear (stress and strain proportional to each other) for small strain, but they may start responding nonlinearly as strain increases. The stress–strain relationship of timber remains essentially linear up to failure, if loaded in a short time. In the case of steel, the stress is proportional to the strain either in tension or compression up to a "yield point," beyond which the stress remains almost constant, whereas the strain increases to the so-called *strain-hardening limit* where stress starts to increase again. The stress–strain relationship of concrete is nonlinear in compression. In tension, its strength is approximately one-tenth of its compressive strength and too small to show on the plot.

EXAMPLE 1.2.1

Calculate the stresses, strains, and deformations of the bars shown in Figure 1.2.13. Assume that Young's modulus is 200 N/mm² in each case.

Solution

Substituting the given values into $\sigma = N/A$, $\varepsilon = \sigma/E$, and $e = \varepsilon.L$, we obtain the following Table 1.2.1, which tells us that a thinner and longer bar tends to elongate more.

TABLE 1.2.1
Solution

	Stress (N/mm²)	Strain (no unit)	Elongation (mm)
(a)	4	0.02	20
(b)	8	0.04	40
(c)	8	0.04	80

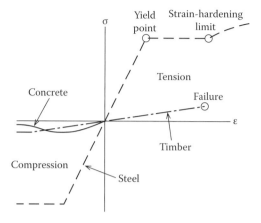

FIGURE 1.2.12 Examples of stress–strain relationships.

EXAMPLE 1.2.2

A force of 500 N is applied to point B as shown in Figure 1.2.14. Calculate the stresses and strains in bars AB and BC. Also calculate the movements (or the *displacements*, to use the technical term) of points B and C.

Solution

Pin support A can carry a horizontal reaction but roller support C cannot. Therefore, the force at point B travels through bar AB to support A, resulting in a tensile axial

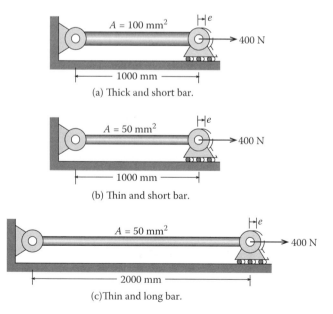

FIGURE 1.2.13 Bass subjected to tensile forces.

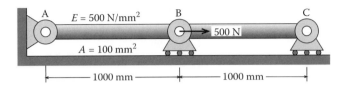

FIGURE 1.2.14 Force on connected bars.

force of 500 N. The stress in bar AB is $\sigma = P/A = 500/100 = 5$ N/mm², and the strain is $\varepsilon = \sigma/E = 5/500 = 0.01$. The axial force, stress, and strain in bar BC are zero. The deformation of bar AB is $e = \varepsilon.L = 10$ mm, and that of bar BC is zero. Because point A cannot move horizontally, points B and C move right by 10 mm (Figure 1.2.15).

EXAMPLE 1.2.3

Three forces, F, F, and $2F$, are applied to a bar as shown in Figure 1.2.16. Find the correct answer among the following values (Table 1.2.2) for the displacement of the bottom end of the bar, if A is the section area, E is Young's modulus, and the self-weight of the bar is negligible.

**TABLE 1.2.2
Options**

1	2	3	4	5
0	$\dfrac{FL}{EA}$	$\dfrac{2FL}{EA}$	$\dfrac{3FL}{EA}$	$\dfrac{4FL}{EA}$

Solution

We first need to evaluate the axial force in the bar. Because there are several external forces, the technique of the imaginary cut (or the "free-body diagram") is useful. With an imaginary cut in the bar as shown in Figure 1.2.17b, we can determine that the axial force between C and D is $2F$. Similarly, Figure 1.2.17c,d give the axial forces in segments BC and AB: F and zero, respectively. The results are summarized in Figure 1.2.17e. Note that the axial force changes at the points where forces are applied. Based on the axial force, we calculate stress, strain, and deformation (or lengthening) as shown in the first row of Table 1.2.3. Adding the deformations for segments AB, BC,

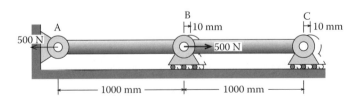

FIGURE 1.2.15 Displacement of points B and C.

Trusses

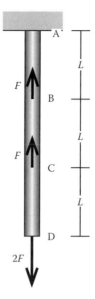

FIGURE 1.2.16 Bar subjected to three forces.

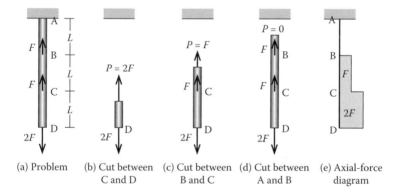

FIGURE 1.2.17 Imaginary cuts resulting in axial-force diagram.

TABLE 1.2.3
Deformation in each part

	Axial Force	Stress	Strain	Deformation
AB	0	0	0	0
BC	F	F/A	F/AE	FL/AE
CD	$2F$	$2F/A$	$2F/AE$	$2FL/AE$

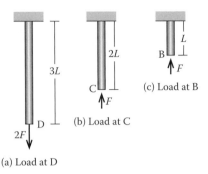

(a) Load at D

(b) Load at C

(c) Load at B

FIGURE 1.2.18 Another approach.

and CD, we obtain the total deformation (or the displacement at point D) of $3FL/AE$. Therefore, the fourth answer is correct.

This problem can be solved using another approach. The force $2F$ at point D elongates the bar by $6FL/AE$ (Figure 1.2.18a). The force F at points C and B shorten the bar by $2FL/AE$ and FL/AE, respectively (Figure 1.2.18b,c). Adding these deformations, we obtain the same result.

$$e = \frac{6FL}{EA} - \frac{2FL}{EA} - \frac{FL}{EA} = \frac{3FL}{EA}$$

Problem: Compute the displacements of points B and C in Example 1.2.3.

Another question in reference to the bar described in Figure 1.2.16: Compute the displacements of points B, C, and D, assuming that an upward force of $2F$ and a downward force of $3F$ are applied at points C and D, respectively.

What Is an Axial Force?

Hi, Joan. I still do not fully understand what an axial force is. What is the difference between an axial force and an external force?

Good question, Jack. An external force is a vector represented by an arrow. It has a magnitude as well as a direction. You see, an external force has a direction such as "to the right" or "down." On the other hand, an axial force is not a vector. You need a pair of arrows with opposite directions to describe an axial force. It represents the magnitude of force with which a member is pulled in tension or pushed in compression. Therefore, we should call it a scalar. Do you like my explanation, Sir?

External force

Axial force

Let me give you another explanation. If a member is pulled in tension, each atom in the member is also pulled in tension by a pair of forces in opposite directions. The axial force is the total of these actions in a sliced section.

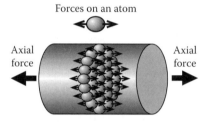

Oh, I see. That is the reason why the axial force has a pair of arrows. Then, how about a reaction? Is it a vector?

Yes, it is, because a reaction has a direction such as "from the wall to the bar." A reaction is similar to an external force in the sense that it is applied to a member from outside.

OK. I see that the pair of an external force and a reaction makes an axial force. But structural mechanics looks too abstract to me. I am losing my confidence.

Do not worry. If you understand the concept of an axial force, the rest is easy. Cheer up.

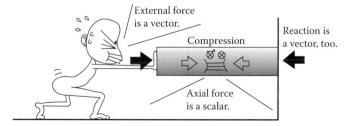

1.3 SYMMETRICAL TRUSS WITH TWO ELEMENTS OF EQUAL SIZE AT 90°

Click the truss shown in Figure 1.3.1a in GOYA-T. The truss you will see consists of two members connected at the top so that the members can rotate as shown in Figure 1.3.1b. This type of connection is called a *pin joint*, or more often, a *joint*, and the point at which two members intersect is called a *node*. The intersecting members are assumed to be connected by a pin perpendicular to the axes of the members, and the contact surfaces of the pin shaft and the members are assumed to have no friction. One of the functions of a pin joint is to prevent forces that may be caused by rotation of the members related to thermal or other effects. The two truss members are connected to the floor using pin supports.

In GOYA-T, the mouse is used to apply a vertical force F_y (up) to the top node. The node moves upward and the members elongate as shown in Figure 1.3.2. To investigate the cause of the elongation, we shall cut the truss member as shown in Figure 1.3.3a. The resulting system, shown in Figure 1.3.3a is called a *free-body diagram*, or simply, a *free body*. From that, we can infer that the truss members are in tension. (We can arrive at the same conclusion by looking at the stretch or elongation of the truss members in Figure 1.3.2.) Because the free body must be in equilibrium, Newton's third law requires that all vectors representing the forces form a closed triangle as shown in Figure 1.3.3b. Note that the tail of each arrow goes to the head of the adjacent arrow. The lengths of the sides of the triangle indicate the magnitudes of the forces P_{AB}, P_{AC}, and F_y. The vertical components of the forces in the truss members (equal to the projections on the vertical of the sides of the triangle representing the member forces P_{AB} and P_{AC}) must balance the applied external force. We use that condition to

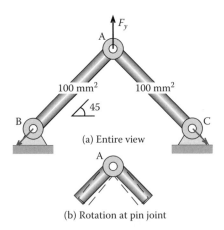

FIGURE 1.3.1 Symmetrical truss with two elements.

Trusses

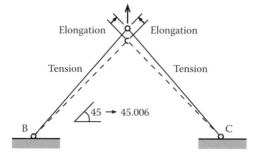

FIGURE 1.3.2 Deformation of the truss.

develop Equation 1.3.1.

$$P_{AB} = P_{AC} = \frac{F_y/2}{\sin 45} = \frac{F_y}{\sqrt{2}} \quad (1.3.1)$$

Having determined that the forces in the truss members are tensile and will result in elongation of the members, we understand why the top node moves up.*

You should also pay attention to the reaction forces at the truss supports. The equilibrium at support C is shown in Figure 1.3.4 (and also in the window of GOYA-T). Note that member AC is in tension. This tensile force causes a reaction of the same magnitude but in the opposite direction at the support.

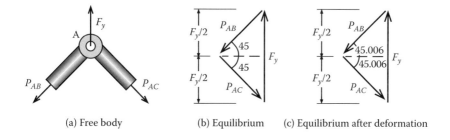

FIGURE 1.3.3 Equilibrium at node A.

---

* *For interested readers*: GOYA-T also shows the displacement of the node. For example, if $F_y = 40$ N, the displacement will be 0.14 mm. This movement widens the angle between the diagonal member and the horizontal from 45° to 45.006° as shown in Figure 1.3.2, slightly changing the equilibrium as shown in Figure 1.3.3c. However, the effect is small enough and can be neglected.

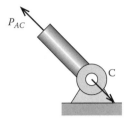

FIGURE 1.3.4 Equilibrium at support C.

Next, apply a horizontal force to node A as shown in Figure 1.3.5a. The resulting equilibrium condition is shown graphically in Figure 1.3.5b. Member AB is pulled in tension by a force

$$P_{AB} = \frac{F_x/2}{\cos 45} = \frac{F_x}{\sqrt{2}} \tag{1.3.2}$$

and member AC is compressed* by a force

$$P_{AC} = -\frac{F_x/2}{\cos 45} = -\frac{F_x}{\sqrt{2}} \tag{1.3.3}$$

The minus sign in the equation indicates that the axial force is compression.† As shown in Figure 1.3.6, the elongation of member AB by the tensile force and the shortening of member AC by the compressive force result in the movement of node A to the right.

---

* If you find it difficult to distinguish whether a member is in tension or compression, assume that both the members are in tension (Figure 1.2a). Then, draw a diagram as shown in Figure 1.2b, where you will find that the tail of the arrow representing P_{AC} touches the tail of F_x. We therefore conclude that the direction of P_{AC} was wrong and that member AC is in compression.

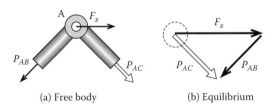

FIGURE 1.2 False equilibrium.

† You may calculate the absolute value first:

$$|P_{AC}| = \frac{F_x/2}{\cos 45} = \frac{F_x}{\sqrt{2}}$$

Then, you need to add a minus sign ($P_{AC} = -F_x/\sqrt{2}$) because the axial force "pushes" the member in the free-body diagram (Figure 1.3.5a).

Trusses

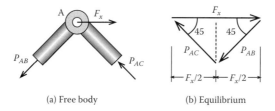

FIGURE 1.3.5 Horizontal force.

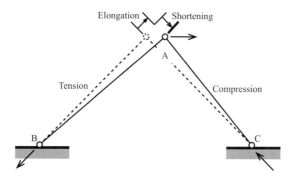

FIGURE 1.3.6 Deformation caused by horizontal force.

EXAMPLE 1.3.1

Calculate the axial forces and the reactions if the truss in Figure 1.3.6 is subjected to an external force of 100 N at node A. The direction of the force is to the right, and makes an angle of 45° with the horizontal.

Solution

Cut the truss (Figure 1.3.7a). The equilibrium defined in Figure 1.3.7b requires that the axial force in member AB is 100 N and that in member AC is zero. Note that node A moves in the direction of the force because member AB elongates and member AC remains at its original length. The tensile axial force of member AB requires a reaction of 100 N at support B. The zero axial force in member AC requires no reaction at support C.

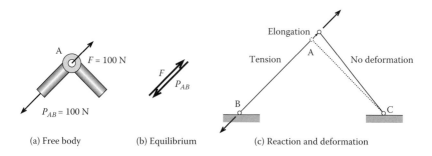

FIGURE 1.3.7 Load at 45° with the horizontal force.

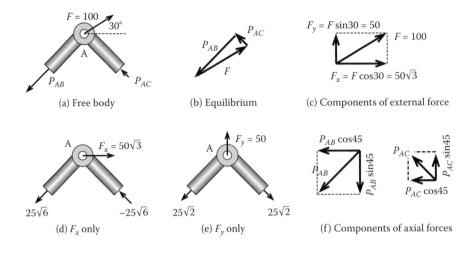

FIGURE 1.3.8 Load at 30° with the horizontal force.

Example 1.3.2

Calculate the axial forces if node A of the truss is subjected to an external force of 100 N pulling right at an angle of 30° with the horizontal (Figure 1.3.8a).

Solution

The force equilibrium is shown graphically in Figure 1.3.8b. To obtain the axial forces, we shall resolve the external force into the horizontal and vertical components as shown in Figure 1.3.8c. To balance the horizontal component, $F_x = 50\sqrt{3}$ N, requires the following axial forces in the truss members (Figure 1.3.8d)*:

$$P_{AB} = \frac{F_x}{\sqrt{2}} = 25\sqrt{6} \text{ N} \qquad P_{AC} = -\frac{F_x}{\sqrt{2}} = -25\sqrt{6} \text{ N}$$

The vertical component, $F_y = 50$ N, requires the following axial forces (Figure 1.3.8e).

$$P_{AB} = P_{AC} = \frac{F_y}{\sqrt{2}} = 25\sqrt{2} \text{ N}$$

Adding the forces P_{AB} and P_{AC}, the following results are obtained:

$$P_{AB} = 25(\sqrt{6} + \sqrt{2}) \approx 97 \text{ N} \qquad P_{AC} = 25(-\sqrt{6} + \sqrt{2}) \approx -26 \text{ N}$$

---

* Recall that the minus sign for P_{AC} simply indicates that the axial force is compression. It does not indicate a change in the direction of the arrow of P_{AC} in Figure 1.3.8d.

Trusses

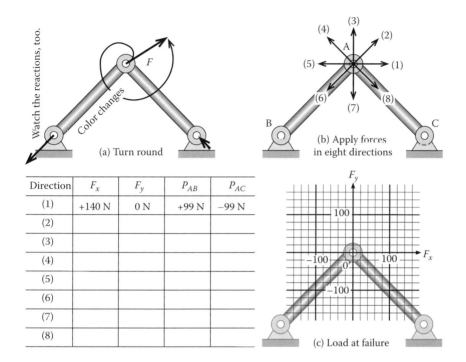

FIGURE 1.3.9 Strength of the truss.

Problem Using GOYA-T

If the axial force exceeds 100 N, the color of each member turns into blue in tension and red in compression, symbolizing tensile or compressive failure of the member. Failure of one of the members means the failure of the truss itself. Click on the node and drag the top node with your mouse as shown in Figure 1.3.9a to see how the axial forces and the reactions vary. Next, apply a force in eight directions successively as shown in Figure 1.3.9b and record the external forces and the axial forces corresponding to failure of the truss in the table included in Figure 1.3.9. Do not expect the axial force to be listed exactly as 100 N. It may be listed, for example, as 99 N because of numerical errors. Plot the results in Figure 1.3.9c to find the weakest direction.

Example 1.3.3

Assume that each truss member cannot resist more than 100 N in tension or compression. Given that a single external force is applied at angle θ with the horizontal at the top node, determine the horizontal, F_x, and the vertical, F_y, components of that force corresponding to the limiting axial force reached in either of the truss members for any θ varying from 0 to 360° (not necessarily at intervals of 45° as shown in Fig 1.3.9b). Plot the relationships with F_x on axis x and F_y on axis y.

22   Understanding Structures: An Introduction to Structural Analysis

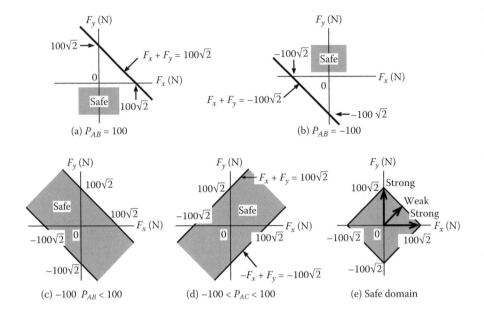

FIGURE 1.3.10 Safe limit of the truss.

Solution

In this example, we generalize the discrete results obtained in the GOYA-T solution in relation to the simple truss shown in Figure 1.3.9b. This time, we work algebraically to define a domain in terms of the vertical and horizontal components of the forces F_y and F_x within which the truss does not fail. The result is shown in Figure 1.3.10e. We reach it by deriving a series of expressions for combinations of F_x and F_y that lead to failure in one of the truss members.

First, we ask what combinations of F_x and F_y would lead to tensile failure. The force in truss member AB is given by adding Equations 1.3.1 and 1.3.2.

$$P_1 = \frac{F_x + F_y}{\sqrt{2}}$$

Tensile failure is assumed to occur if the axial force P_{AB} is 100 N or if

$$\frac{F_x + F_y}{\sqrt{2}} = 100$$

We can rewrite the preceding expression as

$$F_y = 100\sqrt{2} - F_x$$

which plots as a straight line as shown in Figure 1.3.10a. It gives us the limiting combinations of F_x and F_y. Any combination of F_x and F_y to the left of the line is "safe," that is, it does not lead to a tensile force more than 100 N in truss member AB.

Trusses

Similarly, a compressive failure occurs in truss member AB if $P_{AB} = -100$ N or

$$F_y = -100\sqrt{2} - F_x$$

which plots as a straight line in Figure 1.3.10b. This line identifies the limiting combinations of F_x and F_y that may lead to a compressive force of 100 N. Any combination of F_x and F_y to the right of the line is "safe." It does not lead to a compressive force of more than 100 N in truss member AB.

We plot the two lines together in Figure 1.3.10c. The region between the lines contains combinations of F_x and F_y that do not cause the limiting force of 100 N in member AB either in tension or in compression.

We repeat the process for member AC to obtain the two lines in Figure 1.3.10d. These lines bound combinations of F_x and F_y that do not cause the limiting force of 100 N in member AC.

In Figure 1.3.10e, we plot all four lines to define the "safe domain." Combinations of F_x and F_y within the shaded area bounded by the four lines do not lead to forces exceeding 100 N in either member AB or AC. Figure 1.3.10e shows that the truss is strongest if pulled in the x or y direction, and is weakest if pulled at 45°.

Advanced exercise using GOYA-T: You can change the tensile or compressive strength of the material using the window at the lower left corner. If you change the compressive strength to 2 N/mm² while keeping the tensile strength at 1 N/mm², what will the safe domain be?

EXAMPLE 1.3.4

Show the safe domain of the truss in Figure 1.3.11, where each member fails at an axial force of 100 N.

Solution

Denoting the axial forces of the horizontal and vertical members as P_{AB} and P_{AC}, we have $F_x = P_{AB}$ and $F_y = P_{AC}$. Because the upper limit of axial force is 100 N, the allowable range of the external force is

$$|F_x| < 100 \text{ N} \quad \text{and} \quad |F_y| < 100 \text{ N}$$

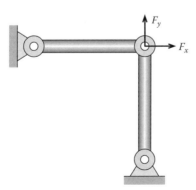

FIGURE 1.3.11 Another truss.

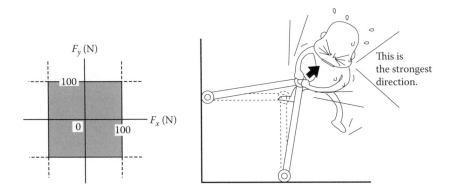

FIGURE 1.3.12 Safe domain.

The shaded area in Figure 1.3.12 shows the safe domain. As long as the combination of F_x and F_y can be plotted within the shaded area, the truss members are not overloaded. Note that the truss in Figure 1.3.11 is identical to that in Figure 1.3.9 except that it is rotated by 45° clockwise. The safe domain in Figure 1.3.12 can also be obtained by rotating Figure 1.3.10e by 45°.

EXAMPLE 1.3.5

We want to design a truss to carry a baby elephant weighing 1000 lbf as shown in Figure 1.3.13. Assume that the tensile strength of the material is 5000 psi. What are the required cross-sectional areas of the truss members?

Solution

The gravity force related to the elephant is 1000 lbf. The axial force in each member is $1000/\sqrt{2} = 500\sqrt{2}$ lbf. Because the stress in the member is given by $\sigma = P/A$, the required cross-sectional area is $A = P/\sigma = 500\sqrt{2}/5000 = 0.1\sqrt{2}$ in.², or approximately 0.14 in.².

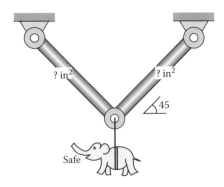

FIGURE 1.3.13 Truss to carry an elephant.

Trusses

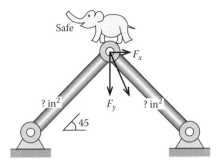

FIGURE 1.3.14 Truss to support an elephant.

Design Your Own Truss (Part 1)

We want to design another truss to support the baby elephant as shown in Figure 1.3.14. Assume that the compressive strength of the material is 1000 psi and the weight of the elephant is 2000 lbf plus any three numbers you may choose (e.g., 2650 lbf). Consider also a lateral seismic force (F_x in the figure) of one half the gravity force. The truss should carry a vertical force of $F_y = 2650$ lbf and a horizontal force of $F_x = \pm 1325$ lbf. What are the required cross-sectional areas of the truss members? Hint: The needed area should be between 2 and 4 in.². (In reality, the truss should be three-dimensional to sustain a wind or seismic force in the direction perpendicular to the paper. We will study such a truss in Section 1.7.)

Coffee Break

Leonardo da Vinci (1452–1519) was the first person to calculate the axial force in a truss. However, he kept his finding in his private notebook and never let the public know. As a result, engineers in the 15th and 16th centuries determined structural member sizes based only on experience, as did the ancient Romans.*

1.4 SYMMETRICAL TRUSS WITH TWO ELEMENTS AT VARIOUS ORIENTATIONS

Start GOYA-T. Select a truss with two members of equal length as shown in Figure 1.4.1a and apply an upward force to node A. The two members elongate, and the node moves upward as shown in Figure 1.4.1b. Cut the truss near node A as shown in Figure 1.4.2a to obtain equilibrium in Figure 1.4.2b leading to

$$P_{AB} = P_{AC} = \frac{F_y/2}{\sin 60} = \frac{F_y}{\sqrt{3}} \quad (1.4.1)$$

---

* Timoshenko, S. P. 1953. *History of Strength of Materials*. McGraw-Hill, New York.

26 Understanding Structures: An Introduction to Structural Analysis

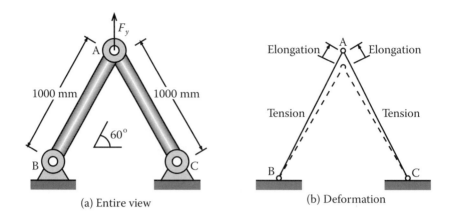

FIGURE 1.4.1 Truss of equilateral triangle.

Next, apply a horizontal force to node A. You will see the truss deflecting as shown in Figure 1.4.3, and equilibrium in the free-body diagram (Figure 1.4.4). The axial force in each member is given by the following equations, in which the minus sign represents compression.

$$P_{AB} = \frac{F_x/2}{\cos 60} = F_x \qquad (1.4.2)$$

$$P_{AC} = -\frac{F_x/2}{\cos 60} = -F_x \qquad (1.4.3)$$

Problem Using GOYA-T

As in the previous section, the color of each member changes if the axial force exceeds 100 N, indicating failure of the member. Drag the mouse spirally as shown in Figure 1.4.5a to see how the axial forces and the reaction vary. Next, apply a force in eight directions as shown in Figure 1.4.5b and record the internal and external forces

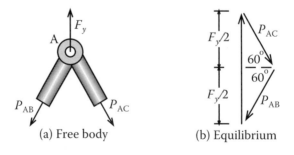

FIGURE 1.4.2 Equilibrium at node A.

Trusses

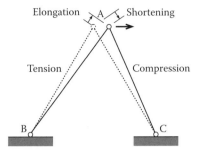

FIGURE 1.4.3 Deformation caused by the horizontal force.

corresponding to the failure of the truss. When you apply the force in the direction indicated as (2), you will find that the node A does not move in the direction of the force but at an angle of 30° with the horizontal as shown in Figure 1.4.5d because the right member AC has no axial force and therefore does not change its length. Thus, the member AC simply rotates around the support C without any change in length.

EXAMPLE 1.4.1

Derive the equations to represent the axial forces in the members of the truss shown in Figure 1.4.5b if an external force is applied successively on the truss in different directions at node A. Then draw a graph to show the safe domain of the truss in the $F_x - F_y$ plane.

Solution

P_{AB} is given by combining Equation 1.4.1 and 1.4.2.

$$P_{AB} = F_x + \frac{F_y}{\sqrt{3}} \tag{1.4.4}$$

P_{AC} is given by combining Equation 1.4.1 and 1.4.3.

$$P_{AC} = -F_x + \frac{F_y}{\sqrt{3}} \tag{1.4.5}$$

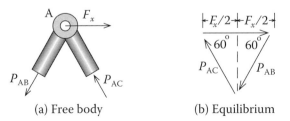

(a) Free body (b) Equilibrium

FIGURE 1.4.4 Equilibrium at node A.

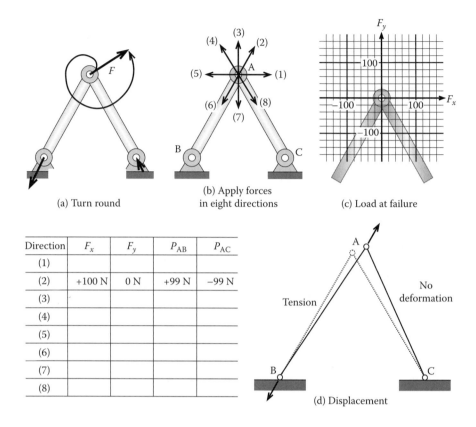

FIGURE 1.4.5 Strength of the truss.

Direction	F_x	F_y	P_{AB}	P_{AC}
(1)				
(2)	+100 N	0 N	+99 N	−99 N
(3)				
(4)				
(5)				
(6)				
(7)				
(8)				

From the conditions $-100 \leq P_{AB} \leq 100$ and $-100 \leq P_{AC} \leq 100$, we obtain the shaded region in Figure 1.4.6a and b, respectively. Combining these figures, we obtain Figure 1.4.6c.

Example 1.4.2

Evaluate the direction in which the truss is most vulnerable to failure and determine the magnitude of the force at which the truss fails. Then calculate the axial force of the members at the failure.

Solution

Focusing on one quadrant of the plot in Figure 1.4.6c, we see in Figure 1.4.7a that the weakest direction is represented by the arrow in the figure (30° from the horizontal and perpendicular to the boundary). The magnitude of the force is

$$F = 100 \cos 30 = 50\sqrt{3} \text{ N}$$

Trusses

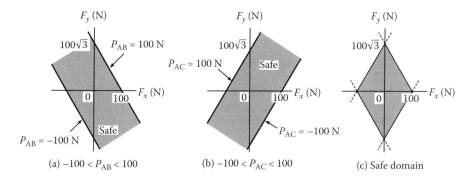

FIGURE 1.4.6 Safe domain of the truss.

The axial forces of the members are determined in Figure 1.4.7b,c:

$$P_{AB} = \frac{F}{\cos 30} = 100 \text{ N} \qquad P_{AC} = -F \tan 30 = -50 \text{ N}$$

EXAMPLE 1.4.3

Draw a graph to show the safe domain for the truss in Figure 1.4.8 assuming that the members fail at 100 N both in tension and compression.

Solution

The axial forces are

$$P_{AB} = \frac{F_x}{2\cos\theta} + \frac{F_y}{2\sin\theta} \qquad (1.4.7)$$

$$P_{AC} = -\frac{F_x}{2\cos\theta} + \frac{F_y}{2\sin\theta} \qquad (1.4.8)$$

From the conditions $-100 \leq P_{AB} \leq 100$ and $-100 \leq P_{AC} \leq 100$, we obtain the shaded region in Figure 1.4.9.

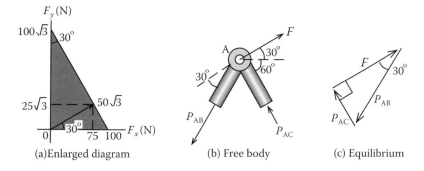

FIGURE 1.4.7 Most vulnerable direction.

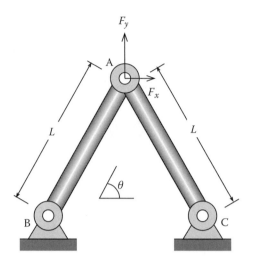

FIGURE 1.4.8 Equilateral truss with angle θ.

Example 1.4.4

Draw a graph to represent the safe domain of the truss shown in Figure 1.4.8 assuming that the members fail at 100 N in tension and at 200 N in compression.

Solution

From the conditions $-200 \leq P_{AB} \leq 100$ and $-200 \leq P_{AC} \leq 100$, we obtain the shaded region in Figure 1.4.10.

Design Your Own Truss (Part 2)

We want to design another truss with members at 60° to one another to support an elephant. Assume that the compressive strength of the material is 1000 psi. The total weight of the elephant is 2000 lbf plus any three numbers you may choose. Consider

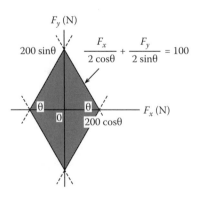

FIGURE 1.4.9 Safe domain (Example 1.4.3).

Trusses

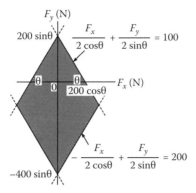

FIGURE 1.4.10 Safe domain. (Example 1.4.8)

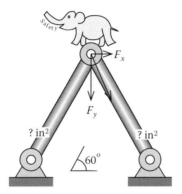

FIGURE 1.4.11 Your truss.

also a lateral seismic force (F_x in Figure 1.4.11) of half the gravity force. How much cross-sectional area is required for each member? (Hint: The needed area will be slightly larger than that in the previous section.)

1.5 UNSYMMETRICAL TRUSS WITH TWO ELEMENTS

Start GOYA-T and click the truss with members having different cross-sectional areas as shown in Figure 1.5.1. Apply an upward force at node A. You will see that the node does not move vertically in line with the applied force but moves to the left as it moves up (Figure 1.5.2).

To understand the reason for this movement, cut the truss as shown in Figure 1.5.3a and consider the equilibrium diagram in Figure 1.5.3b. The axial forces of the members are given as follows:

$$P_{AB} = P_{AC} = \frac{F_y/2}{\sin 60} = \frac{F_y}{\sqrt{3}} \qquad (1.5.1)$$

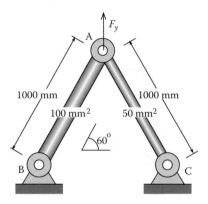

FIGURE 1.5.1 Unsymmetrical truss.

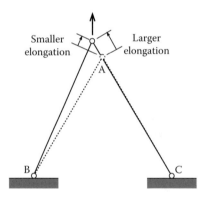

FIGURE 1.5.2 Deformation.

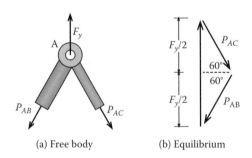

FIGURE 1.5.3 Equilibrium at node A.

Trusses

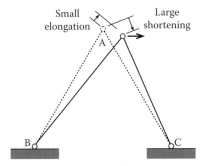

FIGURE 1.5.4 Deformation.

Note that the axial forces are same as those in the previous section though the cross section is different.

However, the stresses in members AB and AC are different because the stress is the axial force divided by the cross-sectional area. The stress in member AC is twice that in member AB, and member AC elongates twice as much as member AB, so that the node moves left as it moves up.

Apply a horizontal force, acting to the right, at node A. The node moves down as it moves to the right (Figure 1.5.4). The absolute magnitudes of the axial forces in the two members are the same as shown in Figure 1.5.5b, but because the cross section of member AC is half that of member AB, AC shortens twice as much as AB elongates.

EXAMPLE 1.5.1

Assume that the strength of the material used for the truss in Figure 1.5.1 is 1 N/mm² both in tension and compression. Show the safe domain of the truss for combinations of forces F_x and F_y.

Solution

The axial forces are given by Equations 1.4.4 and 1.4.5 in the previous section. Because the limits of the axial force of members AB and AC are 100 N and 50 N, respectively, the safe domain for the truss is shown by the shaded area in Figure 1.5.6. At point G ($F_x = 25$ N and $F_y = 75\sqrt{3}$ N), both members fail in tension simultaneously. If you

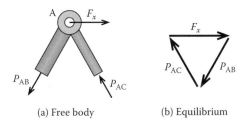

FIGURE 1.5.5 Equilibrium at node A.

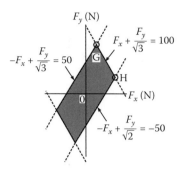

FIGURE 1.5.6 Safe domain.

apply a force in the direction of OG so that $\frac{F_x}{F_y} = \frac{1}{3\sqrt{3}}$, the left and right members elongate equally, and the node moves upward. If you apply a force in the direction of OH so that $F_x/F_y = \sqrt{3}$, the node moves to the right. Apply forces as shown in Figure 1.5.7 to develop a table listing compatible values of F_x, F_y, P_{AB}, and P_{AC} as you did in the previous section, and plot the results in Figure 1.5.6.

We could evaluate the axial forces in all the trusses up to this point considering equilibrium. Such a structure is described as *statically determinate*. Figure 1.5.8 shows another example in which the axial force in member AB is the same as that member BC but the elongation of member AB is more than that of BC. In this case, we can determine the forces in each member using the condition of horizontal equilibrium.

On the other hand, there are structures for which axial forces cannot be evaluated unless the deformation of each member is considered. Such a structure is described as *statically indeterminate*. Figure 1.5.9 shows an example in which the elongations of the two members are the same but the axial forces are different. Equilibrium

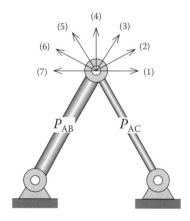

FIGURE 1.5.7 Loading in seven directions.

Trusses

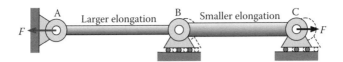

FIGURE 1.5.8 Statically determinate structure.

requires the following equation:

$$F = P_1 + P_2 \tag{1.5.2}$$

Calling the elongation e and the length of the members L, the strain in the members are given by

$$\varepsilon = \frac{e}{L} \tag{1.5.3}$$

Calling the cross-sectional areas of the members A_1 and A_2, and the Young's modulus E, the axial forces are

$$P_1 = A_1\sigma = A_1 E\varepsilon = \frac{A_1 E e}{L} \tag{1.5.4}$$

$$P_2 = A_2\sigma = A_2 E\varepsilon = \frac{A_2 E e}{L} \tag{1.5.5}$$

Eliminating the elongation e from these equations, we obtain

$$P_1 = \frac{A_1}{A_1 + A_2} F \tag{1.5.6}$$

$$P_2 = \frac{A_2}{A_1 + A_2} F \tag{1.5.7}$$

Note that the thicker member carries the larger axial force.

Large structures such as bridges and domes are often designed to be statically determinate to prevent thermal stresses in summer and winter. If the left member in

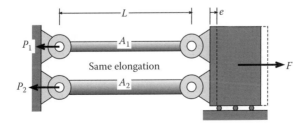

FIGURE 1.5.9 Statically indeterminate structure.

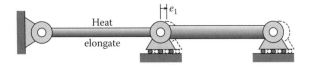

FIGURE 1.5.10 Heat on the left member.

Figure 1.5.10 is heated, the member simply elongates without any additional stresses. On the other hand, if the upper member in Figure 1.5.11 is heated while the lower member is not, the upper member cannot elongate as much as in Figure 1.5.10 ($e_2 < e_1$) because of the restraint by the lower member. As a result, a compressive force appears in the upper member, whereas a tensile force appears in the lower member.

Engineers should be very careful in constructing a statically determinate structure because even a single error may cause a catastrophic collapse. A bad weld in one of the members of the structure in Figure 1.5.8 will cause an immediate collapse, whereas the structure in Figure 1.5.9 might survive a weld failure in one of the members because the remaining member may be able to carry the applied load F.

This book deals only with statically determinate trusses primarily because they are easier to study and provide a good introduction to understanding structural response.

Exercise using GOYA-T: You can find the truss shown in Figure 1.5.12 in GOYA-T. Assume that the material strength is 1 N/mm² both in tension and compression

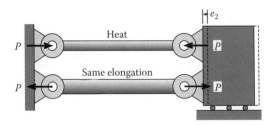

FIGURE 1.5.11 Heat on the upper member.

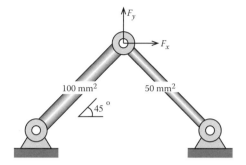

FIGURE 1.5.12 Truss with two elements at different size at 90 degrees.

and draw the safe domain of the truss. After drawing the domain, check it using GOYA-T.

EXAMPLE 1.5.2

You can find the truss shown in Figure 1.5.13a in GOYA-T. Apply an upward force to node A. You will see that the node does not move upward but diagonally as shown in Figure 1.5.13b. Explain the reason.

Solution

Figure 1.5.14 shows equilibrium at node A. All of the external force is carried by the axial force of the vertical member ($P_{AB} = F_y$). Member AC carries no axial force and does not elongate or shorten. Therefore, the length of CA' is the same as that of CA, which is the reason for the diagonal movement of node A (Figure 1.5.13b).

EXAMPLE 1.5.3

Show the safe domain for the truss in Figure 1.5.13a assuming that each member can carry 100 N both in tension and compression.

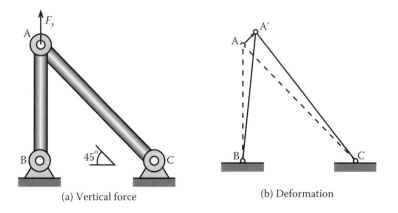

(a) Vertical force

(b) Deformation

FIGURE 1.5.13 Unsymmetrical truss subjected to vertical force.

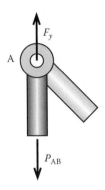

FIGURE 1.5.14 Equilibrium at node A.

Solution

Figure 1.5.15 shows the equilibrium at node A for a horizontal force, F_x. The resulting axial forces are

$$P_1 = F_x \tag{1.5.8}$$

$$P_2 = -\sqrt{2} \cdot F_x \tag{1.5.9}$$

The minus sign in the equation represents that the axial force is compression. The axial forces caused by F_x and F_y are

$$P_1 = F_x + F_y \tag{1.5.10}$$

$$P_2 = -\sqrt{2} \cdot F_x \tag{1.5.11}$$

Because the upper limit of the axial force is 100 N, the safe domain is as shown in Figure 1.5.16.

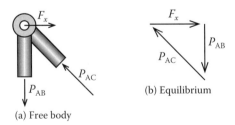

FIGURE 1.5.15 Horizontal force.

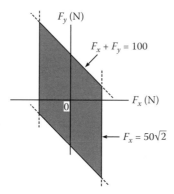

FIGURE 1.5.16 Safe domain.

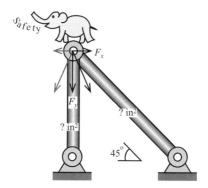

FIGURE 1.5.17 Unsymmetrical truss to support an elephant.

Design Your Own Truss (Part 3)

We want to design another truss to support an elephant as shown in Figure 1.5.17. Assume that the tensile and compressive strengths of the material are 500 and 1000 psi, respectively. The weight of the elephant is 2000 lbf plus any three numbers you may choose. Consider also a seismic force (F_x in the figure) of half the gravity force to the left and the right. How much cross-sectional area is needed for *each* member? (Hint: The needed area will be larger if a seismic force is applied to the left rather than to the right. The area would be between 2.5 and 5 in.²)

1.6 TRUSS WITH THREE MEMBERS

Figure 1.6.1 shows a truss with a pin support at B and a roller support at C. Note that the two supports are connected with a horizontal member. Select this type of truss in GOYA-T and apply an upward force, F_{Ay}, to node A. You will find that the reactions at supports B and C (R_{By} and R_{Cy}) are $F_{Ay}/2$. You will also find that member BC shortens and support C moves to the left.

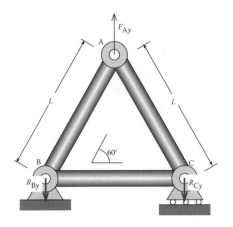

FIGURE 1.6.1 Truss of three members.

Why member BC shortens may be understood as follows. First, look at Figure 1.6.2 showing the free-body diagram of node A. Equilibrium conditions and symmetry help determine the axial forces in the inclined members:

$$P_{AB} = P_{AC} = \frac{F_{Ay}/2}{\sin 60} = \frac{F_{Ay}}{\sqrt{3}} \quad (1.4.1)$$

Second, look at Figure 1.6.3 showing the free-body diagram at support C. Note that the reaction is in the vertical direction. A horizontal component cannot exist because of the roller. The axial force in the bottom member BC is compressive because it has to resist the horizontal component of the axial force in member AC.

$$P_{BC} = -P_{AC} \cos 60 = -\frac{F_{Ay}}{2\sqrt{3}} \quad (1.6.1)$$

The negative sign represents that the axial force in member BC is compressive. Because it is in compression, member BC shortens. The axial forces, P_{AC} and P_{BC}, as well as the reaction R_{Cy}, satisfy the equilibrium shown in Figure 1.6.3b.

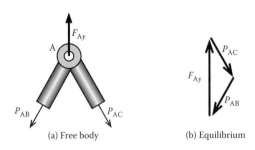

(a) Free body   (b) Equilibrium

FIGURE 1.6.2 Equilibrium at node A.

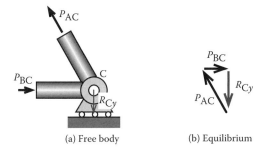

FIGURE 1.6.3 Equilibrium at support C.

Similarly, the equilibrium at support B is shown in Figure 1.6.4.

Apply a force F_{Ax} pulling to the right at node A as shown in Figure 1.6.5. You will find that the horizontal member elongates and support C moves to the right.

Look at Figure 1.6.6 showing equilibrium at node A. It indicates that the axial force in member AB is tensile:

$$P_{AB} = \frac{F_{Ax}/2}{\cos 60} = F_{Ax} \tag{1.6.2}$$

The axial force in member AC is compressive:

$$P_{AC} = -F_{Ax} \tag{1.6.3}$$

Next, look at Figure 1.6.7 showing the free-body diagram at support C. Again, the reaction at C is vertical because of the presence of the roller. The axial force in the bottom member is tensile:

$$P_{BC} = |P_{AC}|\cos 60 = \frac{F_{Ax}}{2} \tag{1.6.4}$$

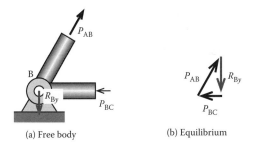

FIGURE 1.6.4 Equilibrium at support B.

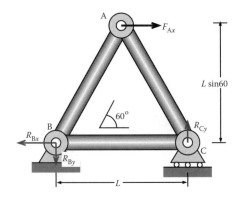

FIGURE 1.6.5 Horizontal force to the truss.

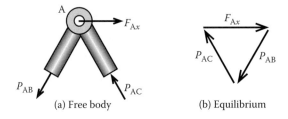

FIGURE 1.6.6 Equilibrium at node A.

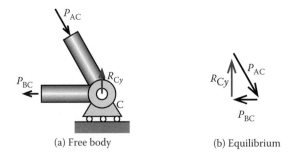

FIGURE 1.6.7 Equilibrium at support C.

Trusses

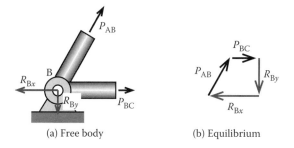

(a) Free body   (b) Equilibrium

FIGURE 1.6.8 Equilibrium at support B.

This tensile force elongates the bottom member, moving node C to the right. The free-body diagram in Figure 1.6.7b can also be used to determine the reaction at C:

$$R_{Cy} = |P_{AC}| \sin 60 = \frac{\sqrt{3}}{2} F_{Ax} \tag{1.6.5}$$

The free-body diagram for support B is shown in Figure 1.6.8. From that we can determine the x- and y-components of the reaction.

$$R_{Bx} = P_{AB} \cos 60 + P_{BC} = \frac{F_{Ax}}{2} + \frac{F_{Ax}}{2} = F_{Ax} \tag{1.6.6}$$

$$R_{By} = P_{AB} \sin 60 = \frac{\sqrt{3}}{2} F_{Ax} \tag{1.6.7}$$

All reaction forces can be and were determined considering equilibrium at the nodes. However, we can calculate the reactions directly if we consider the equilibrium of the truss as a whole. Note that the number of unknown variables in Figure 1.6.5 is three (R_{Bx}, R_{By}, and R_{Cy}). Thus, we need three equations. The first two conditions are:

1. *The sum of the x-components of the forces acting on the structure is zero* ($\Sigma X = 0$). This leads to $-R_{Bx} + F_x = 0$, where R_{Bx} has a negative sign because it goes to the left.
2. *The sum of the y-components of the forces acting on the structure is zero* ($\Sigma Y = 0$). This leads to $R_{By} - R_{Cy} = 0$, where R_{Cy} has a negative sign because it goes downward.

The third condition may be called the principle of the lever, discovered by Archimedes, an ancient Greek philosopher. In the case of the lever scale of Figure 1.6.9, the principle requires $F_1 a_1 = F_2 a_2$. To generalize the principle, we need to introduce a "moment," an action that turns an object around a point as shown in Figure 1.6.10. The moment is defined as follows:

$$(\text{Moment}) = (\text{Force}) \times (\text{Distance}) \tag{1.6.8}$$

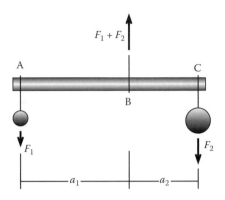

FIGURE 1.6.9 Lever scale.

Equation 1.6.8 is very important. You should understand and remember it. In this book, a clockwise moment is defined as positive and an anticlockwise moment is defined as negative. For an object in equilibrium, the lever principle requires the following:

3. *The sum of the moments acting on the structure is zero* ($\Sigma M = 0$). In the case of the lever shown in Figure 1.6.9, the principle leads to

$$\sum M = -F_1 a_1 + F_2 a_2 = 0 \qquad (1.6.9)$$

where $-F_1 a_1$ and $F_2 a_2$ represent, respectively, the anticlockwise and clockwise moments around point B. In fact, the reference point can be anywhere other than at point B because the lever would not rotate around *any* point. If the reference point is taken as shown in Figure 1.6.11, the principle leads to

$$\sum M = F_1 x - (F_1 + F_2)(x + a_1) + F_2(x + a_1 + a_2) = 0 \qquad (1.6.10)$$

You may find this equation reduces to Equation 1.6.9.

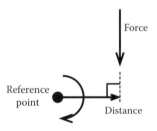

FIGURE 1.6.10 Definition of "moment."

Trusses

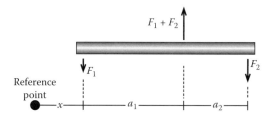

FIGURE 1.6.11 Moment around an arbitrary point.

We return to Figure 1.6.5 and consider the equilibrium of moments around support B.

$$\sum M = F_{Ax} L \sin 60 - R_{Cy} L = 0 \qquad (1.6.11)$$

This results in $R_{Cy} = F_{Ax} \sin 60$, which is equivalent to the solution obtained previously (Equation 1.6.5). But this process makes it much easier to calculate the reactions directly rather than considering the equilibrium of each node.

EXERCISE USING GOYA-T

Assume that each member of the truss shown in Figure 1.6.12 fails at the axial force of 100 N both in tension and compression. Apply forces in seven directions, one at a time, as shown in the figure. Develop a table of the loads, axial forces, and reaction corresponding to the failure of the truss.

EXAMPLE 1.6.1

Plot the safe domain for the truss shown in Figure 1.6.12 assuming that the members fail at 100 N both in tension and compression.

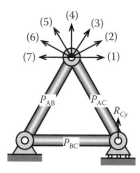

Direction	F_{Ax}	F_{Ay}	P_{AB}	P_{AC}	P_{BC}	R_{Cy}
(1)	+100 N	0 N	+99 N	−99 N	+50 N	+87 N
(2)						
(3)						
(4)						
(5)						
(6)						
(7)						

FIGURE 1.6.12 Apply force in various directions and check the limit.

Solution

The axial forces of the left and right members, P_{AB} and P_{AC}, were determined in Section 1.4.

$$P_{AB} = F_{Ax} + \frac{F_{Ay}}{\sqrt{3}} \quad (1.4.3) \text{ again}$$

$$P_{AC} = -F_{Ax} + \frac{F_{Ay}}{\sqrt{3}} \quad (1.4.4) \text{ again}$$

The axial force of the bottom member, P_{BC}, is determined adding Equation 1.6.1 and 1.6.4.

$$P_{BC} = \frac{F_{Ax}}{2} - \frac{F_{Ay}}{2\sqrt{3}} \quad (1.6.12)$$

Thus, we obtain the shaded "safe" domain in Figure 1.6.13.

Note that the conditions of $P_{BC} = \pm 100$ (the broken lines in the figure) do not affect the shaded region. In fact, Equations 1.6.12 and 1.4.4 lead to $P_{BC} = -P_{AC}/2$, which indicates that the axial force in bottom member BC is always half of that in member AC. Therefore, the truss members need not have equal cross-sectional areas; it is more economical if we make the cross-sectional area of member BC half the cross-sectional area of each of members AB and BC as illustrated in Figure 1.6.14.

EXAMPLE 1.6.2

Calculate the reactions if external forces F_{Ax} and F_{Ay} are applied simultaneously to the truss at node A as shown in Figure 1.6.15. Assume that the reactions in the figure have positive signs.

Solution

The equilibrium conditions $\Sigma X = 0$ and $\Sigma Y = 0$ lead to the following equations.

$$F_{Ax} + R_{Bx} = 0 \quad \text{and} \quad F_{Ay} + R_{By} + R_{Cy} = 0$$

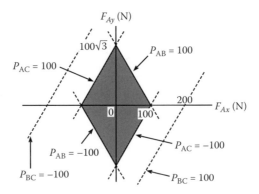

FIGURE 1.6.13 Safe domain.

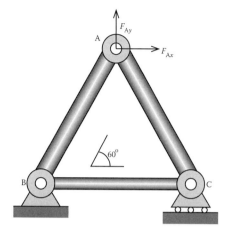

FIGURE 1.6.14 Economical design.

Also, the equilibrium condition $\Sigma M = 0$ around support B leads to

$$F_{Ax}L - R_{Cy}L = 0$$

Thus, we have the following reactions:

$$R_{Bx} = -F_{Ax} \qquad R_{By} = -(F_{Ax} + F_{Ay}) \qquad R_{Cy} = F_{Ax}$$

You can find this type of truss in GOYA-T. Develop a table similar to that in Figure 1.6.12.

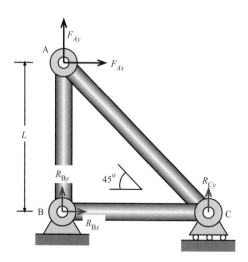

FIGURE 1.6.15 Right-angled truss.

Design Your Own Truss (Part 4)

We want to design another truss. Assume that the tensile and compressive strengths of the material are 500 and 1000 psi, respectively. The weight of the elephant is 2000 lbf plus any three numbers you may choose. Consider also a lateral seismic force (F_x in the Figure 1.6.16) of half the gravity force acting successively in both horizontal directions. What is the minimum cross-sectional area that would be required for each one of the three truss members?

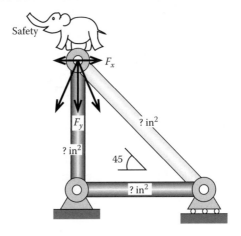

FIGURE 1.6.16 Truss supporting an elephant.

Why Should We Calculate the Reactions?

Sir, I do not understand why we should calculate the reactions.

There are two reasons. First, we need to check whether the ground can sustain the forces from the truss. It would be dangerous if the support fails.

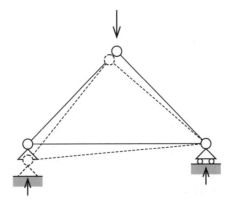

Oh, yes, it is quite obvious. Thanks. Bye.

Wait. There is another important reason. Do you know the reason for determining the axial forces?

Well, to check the safety of the truss. If an axial force exceeds a limit, the truss will fail.

Good. Later on, you will learn to deal with trusses comprising many members. In such trusses, you will have great difficulties if you try to calculate the axial forces using equilibrium conditions at each node. It is much easier to calculate the reactions first. As you know, all the three reactions are obtained from $\Sigma X = \Sigma Y = \Sigma M = 0$, three simultaneous equations with three unknowns.

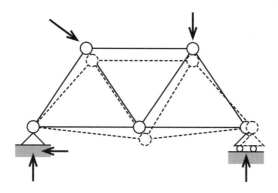

You should be rigorous, Sir. For a truss supported by a pin and a roller, the number of the reactions is three, and therefore $\Sigma X = \Sigma Y = \Sigma M = 0$ is enough. For other cases, however, it may not be enough. For example, a truss supported by two pins has four reactions, and we need another equation.

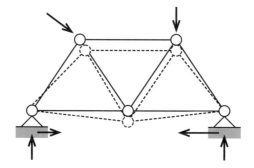

You are right, Joan. But let us ignore such trusses for a while. We may talk about them next year. The important thing is that external forces and reactions always satisfy the condition $\Sigma X = \Sigma Y = \Sigma M = 0$.

Oh, you mean that I can treat the reactions and the external forces equally.

Exactly. Both the reactions and the external forces work from outside the truss. In Section 1.2, you learned that an external force and a reaction work from outside making an axial force in a rod. It is a similar story.

(*For interested readers:* **the moment produced by a force is a vector.**)
Imagine a heavy disc on the x-y plane as shown in Figure 1.6.16. The disk can rotate around the z-axis. An external force **F** (a vector) is applied to the disk at point A, whose location is represented by a vector **a** = OA. Both **a** and **F** are on the x-y plane; θ represents the angle between **a** and **F**. The moment of the force **F** around point O is defined by

$$M = |\mathbf{a}||\mathbf{F}|\sin\theta$$

Note that if the force is in the radial direction (i.e., $\theta = 0$), the force will not turn the disk ($M = 0$). As the angle θ increases up to 90°, the disk tends to turn more easily ($M = |\mathbf{a}||\mathbf{F}|$). Also note that the moment has an axis of rotation (the z-axis), as the screw in Figure 1.6.17. The moment of a force is a *vector* having both a magnitude and a direction. Mathematicians call it a cross-product and express it as follows.

$$\mathbf{M} = \mathbf{a} \times \mathbf{F}$$

The other definition of the product of vectors (a dot product) also plays an important role in mechanics, representing energy or work, which is a scalar. In Figure 1.6.18,

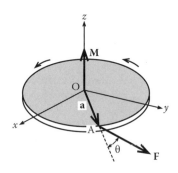

FIGURE 1.6.17 Force on a disk.

FIGURE 1.6.18 Screw.

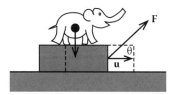

FIGURE 1.6.19 Work on friction energy.

the force **F** and the displacement **u** make the following work, W:

$$W = |\mathbf{F}||\mathbf{u}|\cos\theta$$

1.7 HANDS-ON APPROACH TO TRUSS DESIGN

You need:

1. Four pieces of wood, 2 ft. long, with a square section of 1/8 in. Do not buy balsa wood.
2. A string or thread approximately 7 ft. long.
3. A knife.
4. A scale.
5. A computer or a calculator.
6. Clay to be used as a weight.
7. A plastic bag.
8. A kitchen scale (your instructor may provide one).

In this section, you will make a four-leg truss as shown in Figure 1.7.1. The truss should be designed to collapse if you hang a specific weight at point A. You may choose the weight to be from 2 to 3 lbf.

The elevation of a truss is conceptualized as shown schematically in Figure 1.7.2. The force at the top, F, represents the gravity force caused by the weight. The inclined

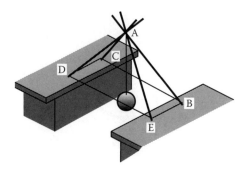

FIGURE 1.7.1 Truss with four legs.

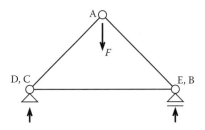

FIGURE 1.7.2 Side view.

members AB, AC, AD, and AE resist compressive forces, and the horizontal members CB and DE resist tensile forces. Therefore, we shall use wood for the compressive members and string for the tensile members.

There are three possible types of collapse modes for the truss:

1. Failure of a connection because of loosening or slip of the string
2. Tensile failure of the string
3. *Buckling* of the wood (a failure mode caused by bending of an axially compressed member as shown in Figure 1.7.3.)

To prevent the first type of collapse, we shall tie the string as tightly as possible.* The second type is not feasible because the string, in tension, is usually stronger than the wood is in compression. The third mode (buckling) can be utilized to realize the desired strength of the truss. The theoretical explanation of the buckling load is beyond the scope of this chapter, and will be given in Chapter 6. In this example, we

---

* http://www.animatedknots.com/cloveboating/.

Trusses

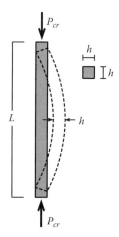

FIGURE 1.7.3 Buckling.

shall use the standard expression engineers employ to determine the buckling load of a concentrically loaded prismatic* column

$$P_{cr} = \frac{\pi^2 E h^4}{12 L^2} \qquad (1.7.1)$$

where E is Young's modulus, h the size of the section, and L the length of the member.

Young's modulus of wood differs, depending on the kinds of trees, but is approximately 1500 ksi. If the wood member measures $1/8 \times 1/8 \times 24$ in., the buckling load is determined to be

$$P_{cr} = \frac{\pi^2 \times 1500 \times 10^3 \times (1/8)^4}{12 \times 24^2} \approx 0.52 \text{ lbf} \approx 8.3 \text{ oz}$$

It is instructive to test whether the calculated value of 8.3 oz is a good measure of the buckling load before we construct the truss. We can check the value using a kitchen scale. Place one end of the wood member, 2 ft. long, on the kitchen scale, and push the other end with your finger slowly until the member buckles.

We do not expect the measured buckling load to be exactly as computed. Young's modulus, the dimensions, and the end conditions of the actual member may be different from those we assumed. However, our calculated answer should be close to

---

* In engineering use, "prismatic" refers to a member that has uniform dimensions and properties along its length. In our application, we are using a wood element having a square section with the same dimensions along its length, and we hope its Young's modulus, E, does not change from end to end.

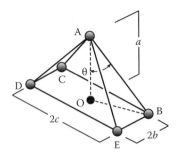

FIGURE 1.7.4 Dimensions.

the measured load unless we have made a critical error in our calculations or in the test.

Having developed confidence in the computed buckling load, how do we relate it to the applied force F on a three-dimensional truss? How do we predict the axial force in the three-dimensional truss? For this purpose, we use Figure 1.7.4, where A, B, C, D, and E are the points at which the wood members are tied. Point O is located directly below A and lies on plane BCDE. The distance between points O and B is

$$|OB| = \sqrt{b^2 + c^2}$$

and the distance between points A and B is

$$L = |AB| = \sqrt{|OA|^2 + |OB|^2} = \sqrt{a^2 + b^2 + c^2}$$

Because of symmetry, we conclude that

$$L = |AB| = |AC| = |AD| = |AE| = \sqrt{a^2 + b^2 + c^2} \tag{1.7.2}$$

This value shall be used to evaluate the buckling load of the inclined members.

Next, we consider the equilibrium at node A as shown in Figure 1.7.5, in which F denotes the external force. From symmetry, we conclude that the axial forces in the inclined members are the same.

$$P_{AB} = P_{AC} = P_{AD} = P_{AE} = P \tag{1.7.3}$$

The sum of the vertical components of the axial forces, $P \cos\theta$, equals the external force, F.

$$F = 4P\cos\theta \tag{1.7.4}$$

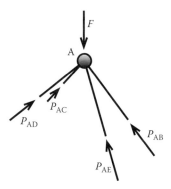

FIGURE 1.7.5 Free-body-diagram.

where θ denotes the angle OAB (Figure 1.7.4):

$$\cos\theta = \frac{|OA|}{|AB|} = \frac{a}{L} \tag{1.7.5}$$

If the axial force is equal to the buckling load, P_{cr}, we have

$$F = \frac{4a}{L} P_{cr} \tag{1.7.6}$$

Let us consider the following two cases.

- The squat truss in Figure 1.7.6, where $a = b = c = 4$ in.
- The tall truss in Figure 1.7.7, where $a = 8$ in. and $b = c = 4$ in.

For both trusses, we use wood members with $h = 1/8$ in. and $E = 1500$ ksi. Equations 1.7.2 and 1.7.6 lead to a relationship between the applied vertical force F and the axial force in each truss member:

- For the squat truss in Figure 1.7.6: $L = \sqrt{4^2 + 4^2 + 4^2} = 6.9$ in. and $F = \frac{4 \times 4}{6.9} P_{cr} = 2.3 P_{cr}$
- For the tall truss in Figure 1.7.7: $L = \sqrt{8^2 + 4^2 + 4^2} = 9.8$ in. and $F = \frac{4 \times 8}{9.8} P_{cr} = 3.3 P_{cr}$

If the buckling load P_{cr} was the same for the two trusses, the squat truss would be weaker than the tall truss. In reality, however, the buckling load is different. It is determined using Equation 1.7.1.

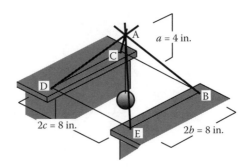

FIGURE 1.7.6 Squat truss.

- For the squat truss in Figure 1.7.6:

$$P_{cr} = \frac{\pi^2 E h^4}{12 L^2} = \frac{3.14^2 \times 1500 \times 10^3 \times (1/8)^4}{12 \times 6.9^2} = 6.3 \text{ lbf}$$

- For the tall truss in Figure 1.7.7:

$$P_{cr} = \frac{\pi^2 E h^4}{12 L^2} = \frac{3.14^2 \times 1500 \times 10^3 \times (1/8)^4}{12 \times 9.8^2} = 3.1 \text{ lbf}$$

The strengths of the trusses are

- For the squat truss in Figure 1.7.6: $F = 2.3 \times 6.3 = 14.5$ lbf
- For the tall truss in Figure 1.7.7: $F = 3.3 \times 3.1 = 10.2$ lbf

So, the taller truss is weaker for the properties we have assumed. On the other hand, if we reduce the dimension a to 0.8 in. and maintain $b = c = 4$ in. (Figure 1.7.8), Equations 1.7.1 and 1.7.6 lead to $P_{cr} = 9.2$ lbf and $F = 0.56 \times 9.2 = 5.2$ lbf, indicating

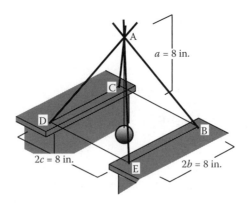

FIGURE 1.7.7 Tall truss.

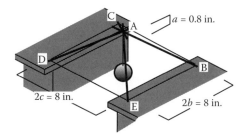

FIGURE 1.7.8 Very shallow truss.

that a very shallow truss with the same base dimensions is even weaker than the tall truss.

If you choose the sizes a, b, and c properly to satisfy Equation 1.7.6, you can design a truss that would fail at a prescribed load, F. We recommend that you use $c = 10$ in. and b equal to a dimension of your choice from 6 to 10 in. so that the class produces various shapes of trusses. The height a shall be determined to be more than 8 in., because a shallow truss is difficult to construct. To determine a, you may use a spreadsheet such as Excel. An example is shown in Figure 1.7.9, where "SQRT" means a square root and "A2^2" indicates the square of the value in cell A2. Increase the value of a from 8 until you get the prescribed load F.

Do not use the ties as shown in Figure 1.7.10a because the wood members bend inward before buckling occurs. Tie them as close to their ends as possible, as shown in Figure 1.7.10b. Small cuts such as those shown in Figure 1.7.10c will help you to tie the string firmly. Tie node B to E first and node A last so that the strings stretch naturally.

At first, the clay weight should be about a half of the target weight. Add to the clay weight until the truss collapses, then weigh the clay. The weight may differ from your expectation because

1. Young's modulus of the wood differs depending on the kind of wood.
2. If you tie node A too firmly, the wood member may not buckle, as shown in Figure 1.7.3, and the buckling load will increase.
3. If the wood member was not straight but curved to start with, the buckling load will decrease.

=3.14^2*1500*10^3*(1/8)^4/(12*D2^2)

	A	B	C	D	E	F
1	a	b	c	L	P_{cr}	F
2	8	8	10	15.1	1.32	2.80
3	9	8	10	15.7	1.23	2.82

=A2 + 1.0   =SQRT(A2^2 + B2^2 + C2^2)

FIGURE 1.7.9 Spreadsheet data.

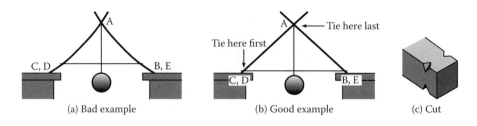

FIGURE 1.7.10 Tie the wood members near their ends first and the top node last.

Write a report on your hands-on experiences:

(a) State the intended and measured strengths of the truss.
(b) State possible reasons for the difference, if any.
(c) Determine the axial forces in the inclined members at collapse using

$$P = \frac{L}{4a} F.$$

(d) Back-figure Young's modulus of the wood ignoring the reasons stated in (b) and using Equation 1.7.1.

1.8 ANALYSIS OF STATICALLY DETERMINATE TRUSSES

Start GOYA-A. It will help you "build" a truss with any arbitrary shape. For example, let us build the one we studied in Section 1.6.

1. Push the "Zoom In" button to have a close-up view of the canvas. Push the "+node" button, and click on the canvas to locate the nodes as shown in Figure 1.8.1. If you click in error, push the "–> node" button and drag the node. You may use the "–node" button to remove a node.

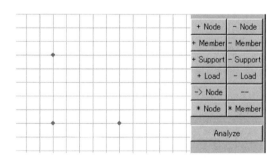

FIGURE 1.8.1 Make nodes.

Trusses

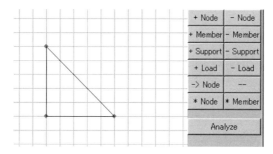

FIGURE 1.8.2 Create members.

2. Push the "+member" button, and drag between nodes to create members as shown in Figure 1.8.2.
3. Push the "+support" button, check "Fix Y" in a new window (Figure 1.8.3) and click the lower left node to make a roller support* that restrains movement in direction Y.
4. Check "Fix X" and click the lower right node to make a pin support that restrains movement in directions X and Y as shown in Figure 1.8.3.
5. Push the "+Load" button to get another window (Figure 1.8.4) and type selected values (in this case, 30 for X and 0 for Y). Click the upper node to show an arrow indicating the force.
6. Push the "Analyze" button to get the result shown in Figure 1.8.4. The numbers indicate the axial forces, positive in tension. Colors also indicate axial forces—blue for tension and red for compression.
7. The size of each square on the screen is assumed to be 10 × 10 mm. Each member is assumed to have a cross-sectional area of 9 mm² and Young's modulus of 10,000 N/mm². The deformation is amplified 100 times. If you move the "Amplification" bar, you can change the amplification factor. If you push the "*Member" button and drag from one end of a member to another, you can see the cross-sectional area of the member and change it if you want.

As another example, let us build the truss shown in Figure 1.8.5. When you draw the members, do not connect A to D directly, but connect AC and CD separately so that your computer distinguishes AC from CD. Figure 1.8.6 shows the resulting truss. Note that the right node (D) moves to the right, not to the left. Can you explain why?

To understand why D moves to the right, let us begin with the reactions. The reaction of the pin support, A, may have x and y components, R_{Ax} and R_{Ay}, whereas that of the roller support has only a y component, R_{Dy}.

---

* Roller supports are represented by (△) rather than (⛋) for brevity.

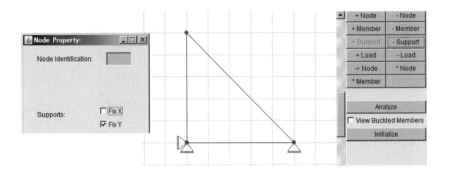

FIGURE 1.8.3 Make supports.

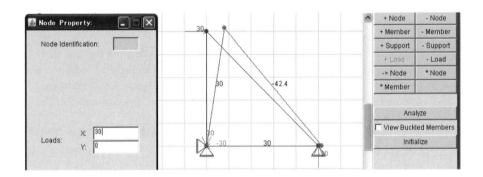

FIGURE 1.8.4 Result.

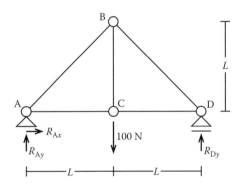

FIGURE 1.8.5 Truss with five members.

Trusses

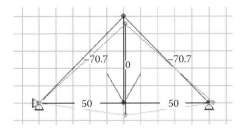

FIGURE 1.8.6 Result of GOYA-A.

First, we consider equilibrium of moments at node A to obtain the equation

$$100 \times L - R_{Dy} \times 2L = 0$$

which leads to $R_{Dy} = 50$ N.

Second, equilibrium in the vertical direction requires

$$-100 + R_{Ay} + R_{Dy} = 0$$

which leads to $R_{Ay} = 50$ N. Third, equilibrium in the horizontal direction leads to $R_{Ax} = 0$. Note that three equilibrium conditions (moment and forces in x- and y-directions) are sufficient to determine the three reactions of a truss supported by a pin and a roller.

Next, we calculate the axial forces considering equilibrium of forces at each node. Let us begin with node A, where only two members intersect. Equilibrium conditions are shown in Figure 1.8.7, leading to $P_{AB} = -50\sqrt{2}$ N (compression) and $P_{AC} = 50$ N (tension). Equilibrium conditions at node C are shown in Figure 1.8.8 leading to $P_{BC} = 50$ N (tension) and $P_{CD} = 50$ N (tension).

We note that both horizontal members AC and CD are in tension. They lengthen. Point A is restrained in the horizontal direction. Point D, restrained only in the vertical direction, has to move to the right.

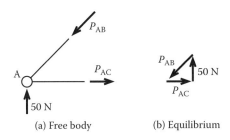

FIGURE 1.8.7 Equilibrium at A.

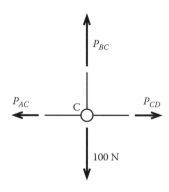

FIGURE 1.8.8 Equilibrium at C.

Example 1.8.1

Calculate the axial force in member BC of the truss shown in Figure 1.8.9.

Solution

For this problem, all we need to do is to consider equilibrium at node C as shown in Figure 1.8.10a. From this we conclude that the axial force in member BC is zero. Incidentally, the reaction at node A is 50 N, and equilibrium at the node is the same as in Figure 1.8.7. The result from GOYA is shown in Figure 1.8.10b, where the horizontal displacement of node D is the same as that in Figure 1.8.6 but the vertical displacement of node C is smaller because member BC does not elongate.

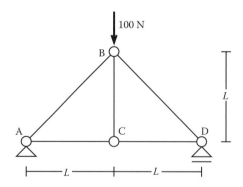

FIGURE 1.8.9 Truss with a load at B.

Trusses

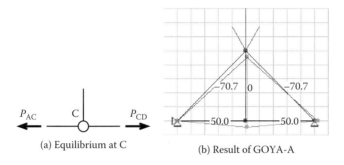

(a) Equilibrium at C

(b) Result of GOYA-A

FIGURE 1.8.10 Solution to Example 1.8.1.

EXAMPLE 1.8.2

Calculate the axial force in member BC in the truss shown in Figure 1.8.11.

Solution

Recognizing symmetry, we obtain the vertical reaction at the support A as 50 N. Equilibrium at node A is shown in Figure 1.8.12, which enables us to determine the axial

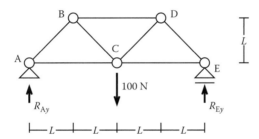

FIGURE 1.8.11 Truss of seven members.

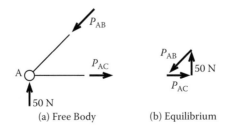

(a) Free Body

(b) Equilibrium

FIGURE 1.8.12 Equilibrium at A.

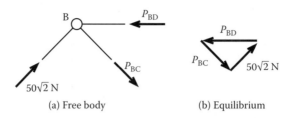

FIGURE 1.8.13 Equilibrium at B.

force in member AB, $P_{AB} = -50\sqrt{2}$ N (compression). Using this value, we consider equilibrium at node B (Figure 1.8.13) and conclude that $P_{BC} = 50\sqrt{2}$ N (tension).

EXAMPLE 1.8.3

Calculate the axial force in member CE of the truss shown in Figure 1.8.14.

Solution

We note the support restraints in Figure 1.8.15 and observe that support E can provide a horizontal reaction. Equilibrium in the horizontal direction leads to $F + R_{Ex} = 0$, which yields $R_{Ex} = -F$, where the negative sign indicates that the reaction is to the left (opposite

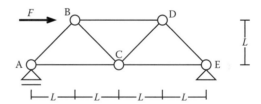

FIGURE 1.8.14 Truss with a horizontal load.

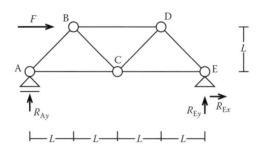

FIGURE 1.8.15 Reactions.

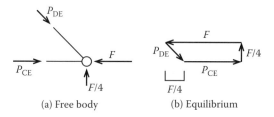

FIGURE 1.8.16 Equilibrium at E.

to the direction shown in Figure 1.8.15). Equilibrium of moment about support A leads to $FL + R_{Ey} \times 4L = 0$, which yields $R_{Ey} = F/4$. Using these values, we set up the force equilibrium at node E (Figure 1.8.16), which yields $P_{CE} = -3P/4$ (compression).

EXERCISE USING GOYA-A

Solve the trusses of Figures 1.8.11 and 1.8.14 using GOYA-A. Sketch the shapes of the trusses before and after deformation using black and red pens, respectively. State the reasons for the following results:

(a) Node B moves to the lower right for the case described in Figure 1.8.11.
(b) Node D moves to the upper right for the case described in Figure 1.8.14.

Up to this point, we isolated each node and considered its equilibrium to determine the forces in the truss members. This procedure, called the *method of joints*, is efficient for simple trusses but not for sophisticated trusses. There is another procedure called the *method of sections*, which makes use of an important condition: *if a structure is in equilibrium, all of its parts must be in equilibrium.* The method is simple: divide a structure into two parts and consider equilibrium in one of the parts. An example follows.

EXAMPLE 1.8.4

Calculate the axial forces in members AB and CD of the truss shown in Figure 1.8.17.

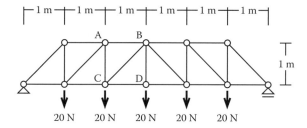

FIGURE 1.8.17 Truss of 21 members.

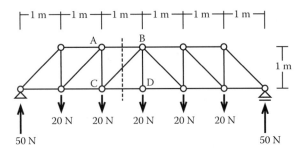

FIGURE 1.8.18 Reactions.

Solution

Note that, for the loading and support conditions given, there is no horizontal reaction at either support. Considering symmetry, we obtain the vertical reaction at each support as 50 N (see Figure 1.8.18). Apply an imaginary cut at the broken line in Figure 1.8.18 and consider the free-body diagram shown in Figure 1.8.19, which includes three unknown axial forces, P_{AB}, P_{CB}, and P_{CD}. Recall that moments must balance around any point. If we consider equilibrium around point C, the contributions of P_{CB} and P_{CD} disappear, and we obtain

$$\sum M = 50 \times 2 - 20 \times 1 + P_{AB} \times 1 = 0 \qquad \text{(Positive clockwise)}$$

which leads to $P_{AB} = -80$ N (compression). Equilibrium in the vertical direction leads to

$$\sum Y = 50 - 20 - 20 + P_{CB} \sin 45 = 0 \qquad \text{(Positive upward)}$$

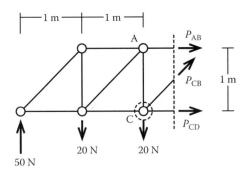

FIGURE 1.8.19 Free body.

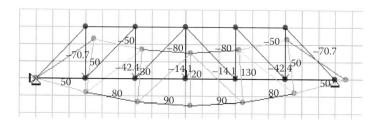

FIGURE 1.8.20 Result of GOYA-A.

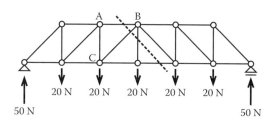

FIGURE 1.8.21 Do not cut a section similar to the one shown!

which yields $P_{CB} = -10\sqrt{2}$ N (compression). Equilibrium in the horizontal direction leads to

$$\sum X = P_{AB} + P_{CB} \sin 45 + P_{CD} = 0 \qquad \text{(Positive right)}$$

which yields $P_{CD} = 90$ N (tension). Results obtained using GOYA-A are shown in Figure 1.8.20. Note that all bottom members are in tension and all top members are in compression. The elongation of the bottom chord (bottom members) and the shortening of the top chord result in downward deflection of the truss.

Note that we can have only three equations based on the conditions of equilibrium at a section: $\Sigma X = 0$, $\Sigma Y = 0$, and $\Sigma M = 0$. Thus, we should "cut" a truss so that the section does not intersect more than three members (Figure 1.8.21).

EXERCISE 1.8.5

Calculate the axial force in member AB in the truss (Figure 1.8.18) for the following cases using the method of sections.

(a) Load of 100 N is applied at node D (and no load at the other nodes).
(b) Load of 100 N is applied at node C (and no load at the other nodes).

Determine the required forces using GOYA-A. Sketch the shapes of the trusses before and after deformation, using black and red pens, respectively. State the difference between these results.

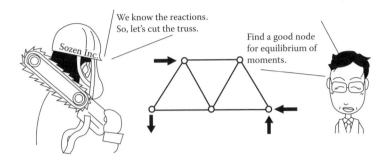

FIGURE 1.8.22 Method of sections.

The method of sections is summarized in Figure 1.8.22:

1. Compute the reactions using $\Sigma X = \Sigma Y = \Sigma M = 0$.
2. Cut the truss.
3. Then, find a good node to set up the equation defining equilibrium of moments

Even if you cannot find a good node do not worry. As long as the cut section does not include more than three unknown axial forces, you can write three equations ($\Sigma X = \Sigma Y = \Sigma M = 0$), which enable you to obtain the solutions.

1.9 STABLE TRUSSES

As stated in Section 1.1, trusses were invented to support heavy loads using small amounts of material. However, if its design is incorrect, a truss can fail at a load smaller than that for which it was proportioned.

An example is shown in Figure 1.9.1a. If you push node A or C, the truss will easily deform to the shape indicated by the broken lines. Such a truss is called an *unstable truss*. Trusses in Figure 1.9.1b–d are other examples. We shall propose a

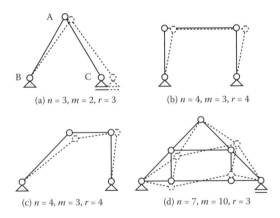

FIGURE 1.9.1 Unstable trusses with $f = 1$.

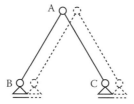

FIGURE 1.9.2 Unstable truss with $f = 2$.

truss contest in Section 1.10. When you take on that challenge, you should be careful about the stability of the truss. If you try to analyze an unstable truss in GOYA-A, your computer will sound an alarm.

In the 19th century, a German mathematician, A. F. Möbius (well known for the "Möbius strip"), found that a truss is unstable if the freedom of motion, f in the following equation, is larger than zero.

$$f = 2n - m - r \tag{1.9.1}$$

where n is the number of nodes,* m is the number of members, and r is the number of restraints imposed by the supports (pins and rollers). For example, r for the truss in Fig. 1.9.1a is 3 because the pin support restrains node B from moving in the x- and y-directions, and the roller support restrains node C from moving in the y-direction. The freedom of motion, determined from Equation 1.9.1 for the truss in Figure 1.9.1a, is $f = 2 \times 3 - 2 - 3 = 1 > 0$. It is unstable. The trusses in Figure 1.9.1b–d also have $f = 1$ and are unstable.

The "degrees of freedom" actually represent the possible motions of the truss nodes that do not require deformation (lengthening or shortening) of its members. In Figure 1.9.1a, the movement of node A is related to that of node C. In other words, the truss can move in a unique fashion. The degree of freedom would be one. If you change support B from a pin to a roller, the degree of freedom would be two. Then, the truss can move as indicated in Figure 1.9.2 without any change in shape. In any case, both trusses (Figures 1.9.1a and 1.9.2) cannot resist a load. If you change support C in Figure 1.9.1a from a roller to a pin (i.e., increase r from 3 to 4 as shown in Figure 1.9.3a), we have $f = 0$ and the truss is stable. The trusses in Figure 1.9.3b–d have one more member than those in Figure 1.9.2 and are stable.

The value of f (Equation 1.9.1) may be obtained in each case as follows:

(a) Each node has the freedom to move in directions x and y. Thus, n nodes provide $2n$ degrees of freedom. The number of degrees of freedom is $2n$.
(b) Each member maintains the distance between two nodes constant and decreases the degrees of freedom by one. The number of degrees of freedom becomes $2n - m$.

---

* A node of a truss is defined as the point where members intersect or are supported.

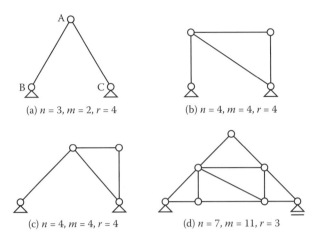

FIGURE 1.9.3 Statically determinate trusses ($f = 0$).

(c) Each restraint decreases the degree of freedom by one. Thus, r restraints decrease r degrees of freedom. The number of degrees of freedom becomes $2n - m - r$ as stated in Equation 1.9.1.

Although a truss with $f > 0$ is unstable, a truss with $f \leq 0$ is not necessarily stable (converse statements are not always true). For example, the truss in Figure 1.9.4 has $f = 0$ but is unstable. Note that the bottom member connecting the pin supports does not work at all.

There is another important rule: the number of degrees of freedom f of a statically determinate truss is zero. Recall the definition of a statically determinate truss: the axial force in each member can be calculated considering equilibrium only. The unknowns in a truss are the axial forces, m, and the reactions, r, adding to $(m + r)$. On the other hand, equilibrium at each node in x- and y-directions leads to two equations or a total of $2n$ equations. In the case of a statically determinate truss, the number of unknowns $(m + r)$ must be equal to the number of equations ($2n$), so that $f = 2n - m - r = 0$.

If you increase the number of members of the trusses in Figure 1.9.3 as shown in Figure 1.9.5, we have more unknowns than available equations. With the added members, the trusses are indeterminate.

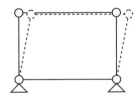

FIGURE 1.9.4 Unstable truss with $f = 0$.

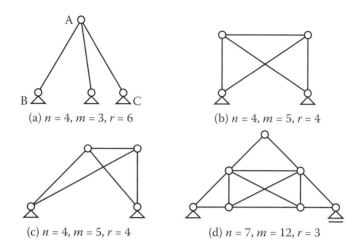

FIGURE 1.9.5 Statically indeterminate trusses ($f = -1$).

FIGURE 1.9.6 Three-dimensional truss ($f = 0$).

For three-dimensional trusses, equilibrium at each node in directions x, y, and z leads to three equations, and the degree of freedom is given by

$$f = 3n - m - r \qquad (1.9.2)$$

For example, the value of r for the truss in Figure 1.9.6, supported by three pins is 9, and its degree of freedom is $f = 3 \times 4 - 3 - 9 = 0$. It is statically determinate. Two types of roller supports can be used in a three-dimensional truss: the one shown in Figure 1.9.7a provides unrestrained movement in two directions and provides a

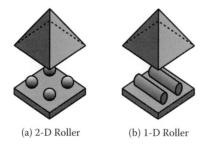

FIGURE 1.9.7 Three-dimensional rollers.

restraint in the vertical direction only ($r = 1$), whereas the other roller support in Figure 1.9.7b provides unrestrained movement in one direction but restrains movement in two directions ($r = 2$).

1.10 BUILDING A TRUSS

CONTEST USING GOYA-A

First, we shall compete for the strongest truss using GOYA-A. Assume the size of each square on the GOYA-A screen to be 10×10 mm.

To "build" your truss, you must satisfy the following five requirements.

1. The truss shall be supported by a roller at one end and a pin at the other end over a span of 200 mm.
2. The load shall be applied at the highest point in the center of the span.
3. The truss shall be statically determinate ($f = 2n - m - r = 0$).
4. The number of members shall not exceed 10.
5. Each compressive member shall have a square section of 3 mm × 3 mm and a Young's modulus of 10,000 N/mm². Axial force in each compressive member shall not exceed the buckling load, P_{cr} in Equation 1.7.1. Members shall be assumed not to be vulnerable to failure in tension.

An example of such a truss is shown in Figure 1.10.1, where the solid lines with numbers (1)–(6) represent the compressive members. After building a truss, check the box "View Buckled Members." Buckled members are shown in red. If all the members are black, the truss is safe.

CONTEST USING WOOD AND STRING

Let us make an actual truss that meets the following five requirements (to date, the record for the maximum load carried has been 16 lbf):

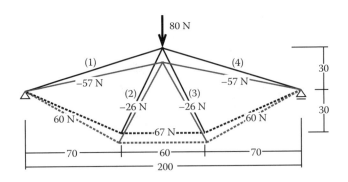

FIGURE 1.10.1 Example.

FIGURE 1.10.2 Winner in year 2004 (M. Hirabayasi).

1. The truss shall be placed on two desks at 2 ft. from one another. (To make certain that it does not slip off one of the desks, the truss span should be approximately 2 ft. 2 in.)
2. The material shall comprise four pieces of wood with a square section (1/8 in. × 1/8 in.) and a length of 2 ft, and a string. You may use as much string as you wish. You cannot use other materials such as adhesives.
3. The only tool you may use is a knife.
4. Before the test, you shall submit a drawing of the truss and a bill of material (lengths of wood pieces used).
5. The load should be applied at the highest point in the center of the span using a string.

Figure 1.10.2 shows the winner of the contest in 2004.

1.11 PROBLEMS

1.1 Calculate the axial forces in members AB and AC in Figure 1.11.1 (P_{AB} and P_{AC}), and select the set closest to your answer from Table 1.11.1.

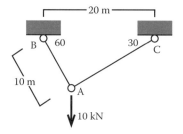

	P_{AB}	P_{AC}
1	5 kN	9 kN
2	6 kN	8 kN
3	7 kN	7 kN
4	8 kN	6 kN
5	9 kN	5 kN

Table 1-11-1

FIGURE 1.11.1 Truss of two members.

1.2 Calculate the elongations of members AB and AC in Figure 1.11.1 (e_{AB} and e_{AC}) assuming that the cross-sectional area of each member is 100 mm² and Young's modulus is 200 kN/mm². Select the set closest to your answer from Table 1.11.2.

TABLE 1.11.2

	e_{AB} (mm)	e_{AC} (mm)
1	2	6
2	3	5
3	4	4
4	5	3
5	6	2

1.3 Calculate the maximum force at node A in Figure 1.11.1 assuming that the tensile strength of the material is 200 N/mm². Select the value closest to your answer from Table 1.11.3.

TABLE 1.11.3

(kN)

1	8
2	13
3	18
4	23
5	28

1.4 Calculate the axial forces in members AC and CE in Figure 1.11.2 (P_{AC} and P_{CE}), and select the set closest to your answer from Table 1.11.4.

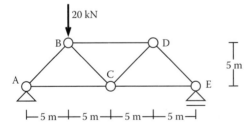

Table 1-11-4

	P_{AB}	P_{AC}
1	15 kN	5 kN
2	14 kN	6 kN
3	13 kN	7 kN
4	12 kN	8 kN
5	11 kN	9 kN

FIGURE 1.11.2 Truss of seven members.

1.5 Calculate the horizontal displacement of the roller support in Figure 1.11.2 assuming that the cross-sectional area of each member is 100 mm² and Young's modulus is 200 kN/mm². Select the value closest to your answer from Table 1.11.5.

TABLE1.11.5

1	2 mm
2	4 mm
3	6 mm
4	8 mm
5	10 mm

1.6 Choose the correct set of axial forces in members AB, BE, DE, and FG in Figure 1.11.3 from those listed in Table 1.11.6, where T, C, and 0 represent tensile, compressive, and zero axial-forces, respectively.

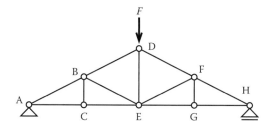

Table 1-11-6

	AB	BE	DE	FG
1	T	0	0	0
2	C	C	C	0
3	C	T	T	T
4	C	0	0	0
5	T	C	T	C

FIGURE 1.11.3 Root truss supporting a force.

1.7 Choose the correct set of axial forces in members AB, BE, DE, and FG in Figure 1.11.4 from those listed in Table 1.11.7, where T, C, and 0 represent tensile, compressive, and zero axial-forces, respectively.

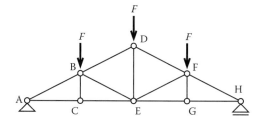

Table 1-11-7

	AB	BE	DE	FG
1	T	C	T	0
2	C	0	C	C
3	C	T	T	T
4	C	C	T	0
5	T	C	T	C

FIGURE 1.11.4 Root truss supporting three forces.

1.8 Calculate the axial force in member AB in Figure 1.11.5 and select the correct answer from Table 1.11.8.

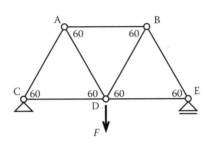

Table 1-11-8

1	$+\sqrt{3}F$
2	$-\sqrt{3}F$
3	$+\dfrac{1}{\sqrt{3}}F$
4	$-\dfrac{1}{\sqrt{3}}F$
5	$+\dfrac{1}{2\sqrt{3}}F$

FIGURE 1.11.5 Trapezoidal truss.

1.9 Calculate the axial force in member AB in Figure 1.11.6, and select the correct answer from Table 1.11.9.

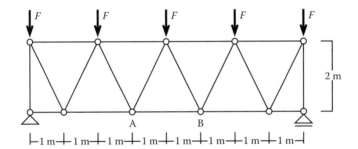

Table 1-11-9

1	$2F$
2	$3F$
3	$4F$
4	$5F$
5	$6F$

FIGURE 1.11.6 Rectangular truss.

1.10 Calculate the axial forces in members A through E in Figure 1.11.7. Which listing for axial force in Table 1.11.10 is incorrect?

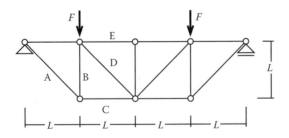

Table 1-11-10

	Member	Axial force
1	A	$+\sqrt{2}F$
2	B	$-F$
3	C	$+F$
4	D	$-\sqrt{2}F$
5	E	$-F$

FIGURE 1.11.7 Truss hanging down from supports.

Trusses

1.11 Calculate the axial force in member AB in Figure 1.11.8 and select the correct answer from Table 1.11.11.

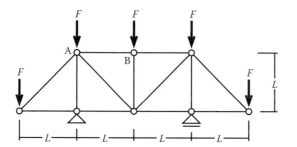

Table 1-11-11

1	0
2	−2.5 F
3	+0.5 F
4	+2.5 F
5	+3.5 F

FIGURE 1.11.8 Overhanging truss.

1.12 Calculate the axial force in member AB in Figure 1.11.9 and select the correct answer from Table 1.11.12.

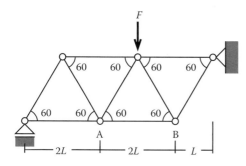

Table 1-11-12

1	$+\dfrac{9}{5\sqrt{3}}F$
2	$+\dfrac{6}{5\sqrt{3}}F$
3	0
4	$-\dfrac{9}{5\sqrt{3}}F$
5	$-\sqrt{3}\,F$

FIGURE 1.11.9 Unsymmetrical truss with a horizontal roller.

1.13 Calculate the axial force in member AB in Figure 1.11.10 and select the correct answer from Table 1.11.13.

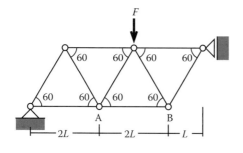

Table 1-11-13

1	$+\dfrac{9}{5\sqrt{3}}F$
2	$+\dfrac{6}{5\sqrt{3}}F$
3	0
4	$-\dfrac{9}{5\sqrt{3}}F$
5	$-\sqrt{3}\,F$

FIGURE 1.11.10 Unsymmetrical truss with a vertical roller.

1.14 Calculate the axial force in member AB in Figure 1.11.11 and select the correct answer from Table 1.11.14.

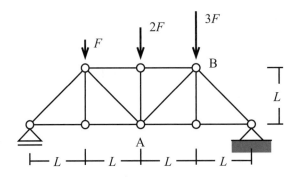

FIGURE 1.11.11 Unsymmetrical loads.

2 Moments and Deflections in Cantilever Beams

2.1 WHAT IS A BENDING MOMENT?

Chapter 1 dealt with trusses. Trusses are used mainly for large structures such as bridges and domes. In this chapter, we shall study beams which are used not only in bridges but also in buildings. A beam is a horizontal member supporting a roof or a floor as shown in Figure 2.1.1a. A beam that projects out from a support and has a free (unsupported) end is called a 'cantilever beam.' To understand the phenomenon of bending of beams in simple steps, we shall start by studying strength and deformation properties of a cantilever beam embedded in a rigid wall (Figure 2.1.1b).

Start 'GOYA-C' to reach the window in Figure 2.1.2 showing a cantilever beam with a load of 10 N at its free end. You will find a sliding bar titled 'Load Magnitude' at the bottom of the window. Using your pointer, move the bar to the left to obtain −10 N. The beam will deflect down as shown in Figure 2.1.3.

How does the beam deflect if it is subjected to the set of forces with the same magnitude (Figure 2.1.4a)? Which of the four deflected shapes in Figures 2.1.4b through e would you guess is the correct one? To check your guess, do the following.

1. Click the 'Add Load' button to create another load pushing up at the end of the beam. The deflection will disappear (Figure 2.1.5).
2. Click on the downward load and drag it to mid-span (Figure 2.1.6).

The entire beam deflects up!

Let us now investigate in detail how the beam bends. Click the 'Zoom in' button and type "4" in the text field titled 'Amplification' to get the image shown in Figure 2.1.7. You will find that the yellow segment rotates and moves up and that the top of the yellow segment shortens while the bottom lengthens.

Click the 'Zoom out' button and type "1" in the 'Amplification' text-field. Look at the window in the upper right (Figure 2.1.8) showing the deformation of the highlighted (yellow) segment. This segment has been removed from the beam and rotated so that its y-axis is vertical. Also, the flexural deformations (the shortening of the top and lengthening of the bottom) have been magnified by a factor of ten.

Why does the top of the segment shorten while the bottom lengthens? The answer can be obtained if you check the 'Free-body Diagram' item in the bottom right to obtain the image shown in Figure 2.1.9. GOYA-C shows a free-body diagram of the beam to the left of the broken line in Figure 2.1.10a. The two external forces tend to rotate the beam to the left of the cut clockwise (Figure 2.1.10b). To prevent the rotation, there should be a counterclockwise action M to the right of the highlighted segment (Figure 2.1.10c). This action is called a *'bending moment.'* The moment is

80   Understanding Structures: An Introduction to Structural Analysis

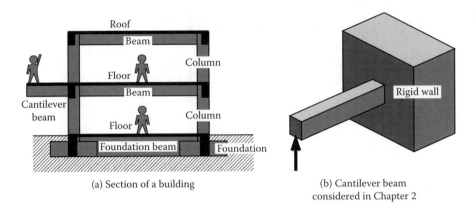

FIGURE 2.1.1 Cantilever beam.

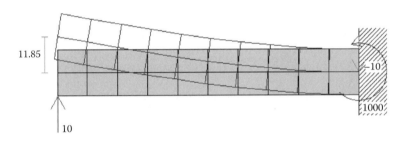

FIGURE 2.1.2 Deflected shape of cantilever as seen in first window of GOYA-C.

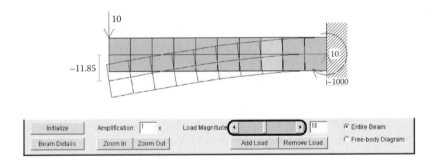

FIGURE 2.1.3 Deflected shape for a load of −10 N applied at free end of a cantilever beam.

Moments and Deflections in Cantilever Beams

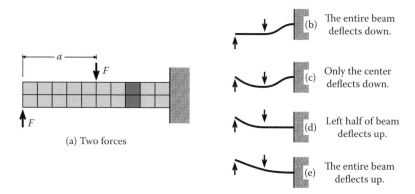

FIGURE 2.1.4 Quiz: a cantilever beam with two forces.

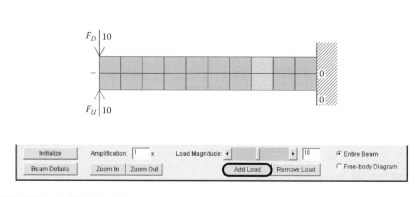

FIGURE 2.1.5 Click 'Add load' button.

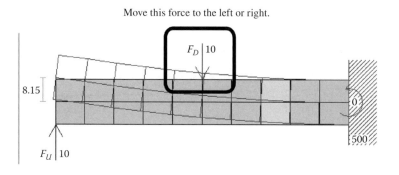

FIGURE 2.1.6 Drag the load to mid-span.

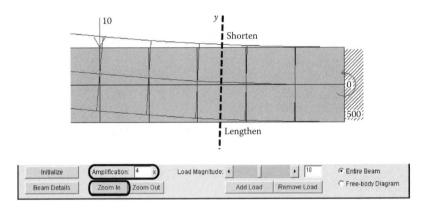

FIGURE 2.1.7 Click the 'Zoom in' button and type "4" in the 'Amplification' text-field.

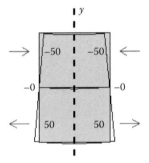

FIGURE 2.1.8 Flexural deformation.

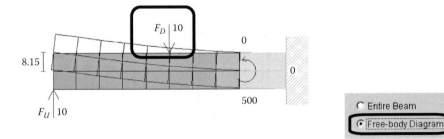

FIGURE 2.1.9 Free-body diagram to the left of the highlighted segment.

Moments and Deflections in Cantilever Beams

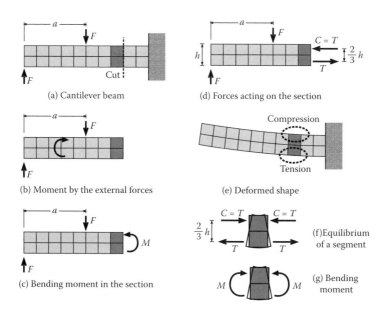

FIGURE 2.1.10 Cut the beam.

defined as a force multiplied by a distance (see Section 1.6). The magnitude of the bending moment is given by Equation 2.1.1.

$$M = F.a \tag{2.1.1}$$

In GOYA-C, the width of each segment is 10 mm, so that the length between the forces is $a = 50$ mm (Figure 2.1.9). From Equation 2.1.1, the moment M is 500 N-mm ($M = F.a = 10 \times 50 = 500$ N-mm). Move the location of the forces to the left or to the right to obtain your own perspective of the variations in bending moment and deflected shape of the beam. Recall that the beam deflection increased when you moved the force F_D to the right in Figure 2.1.6, suggesting that an increase of the distance a in Equation 2.1.1 results in a larger bending moment and larger deflection. Using GOYA-C, test to find out if this suggestion is correct.

Figure 2.1.10c shows a free-body diagram of the beam with the moment M required to balance the moment generated as the product of F and a. The moment M is essentially a *couple* as illustrated in Figure 2.1.10d. It comprises two equal and opposite forces at a distance $(2/3)h$ from one another, where h is the beam height.[*]

---

[*] Strictly speaking, tensile and compressive forces acting on the beam cross section vary linearly over the height of this rectangular cross section starting from zero at mid-height and reaching maxima at the extreme fibers in tension and compression (Figure 2.1.11a). The distribution of the bending stresses will be described in Section 2.4. In this section, these distributed forces are lumped in concentrated forces shown in Figure 2.1.11b for simplicity. The lumped forces are assumed to act at the centroids of the distributed forces, so that the distance between the concentrated forces is two thirds of the height of the beam, $(2/3)h$. The height of the beam is selected to be $h = 15$ mm for the particular beam we are considering. Substituting $F = 10$ N, $a = 50$ mm and $h = 15$ mm into $(2/3)T.h = F.a$, we have $T = 50$ N as indicated in Figure 2.1.8. In a cross section subjected to bending only, if the lumped tensile force is 50 N, the lumped compressive force is also -50 N. One has to balance the other.

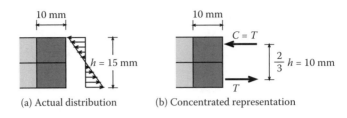

(a) Actual distribution   (b) Concentrated representation

FIGURE 2.1.11 Forces acting on the highlighted segment.

Equilibrium in the horizontal direction requires $T = C$. The product $(2/3)\,hT$ or $(2/3)\,hC$ is equal to M.

The forces, T and C, cause tensile and compressive deformations of the beam (Figure 2.1.10e). To understand their effect, cut the beam again at the left end of the yellow segment as shown in Figure 2.1.10f, where equilibrium in the horizontal direction requires the same magnitude of forces at both faces of the segment. These pairs of tensile and compressive forces create moments causing the flexural deformation (i.e., the lengthening at the top and the shortening of the bottom). The moments are represented by round (bent) arrows as shown in Figure 2.1.10g.

Equilibrium of a beam is similar to that of a truss. Look at the truss in Figure 2.1.12a. It is connected to the wall at its right end and is analogous to the cantilever beam shown in Figure 2.1.12b. Cutting the truss as shown in Figure 2.1.12c

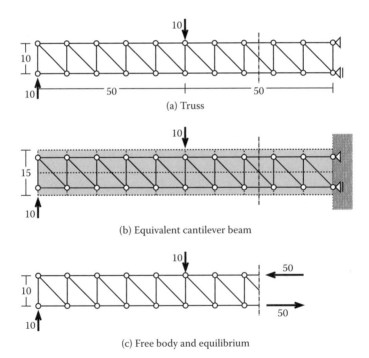

FIGURE 2.1.12 A cantilever truss.

Moments and Deflections in Cantilever Beams

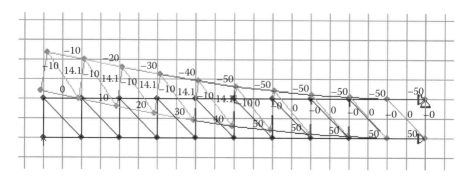

FIGURE 2.1.13 GOYA-A solution for a cantilever truss.

and considering moment equilibrium, you will find that the top chord carries the tensile force of 50 N and the bottom chord the compressive force of 50 N. This is equivalent to the equilibrium condition depicted in Figure 2.1.10d.

Figure 2.1.13 shows the results of the analysis of the truss in Figure 2.1.12a obtained using GOYA-A. The deflected shape of the truss is also similar to that of the cantilever beam.

EXPERIMENT

Make a cantilever beam using a plastic ruler as shown in Figure 2.1.14, and apply a pair of forces using your thumb and little finger. If you can make the magnitude of the forces equal, the ruler will bend up.

EXERCISE

Compare Figures 2.1.15a and b. Which beam has larger bending moment at the highlighted segment? Which beam deflects more?

Answer

The bending moments are the same because the distances between the vertical forces are the same. However, because beam (a) has a longer deformed region than beam (b) does, the deflection of beam (a) is more than that of beam (b).

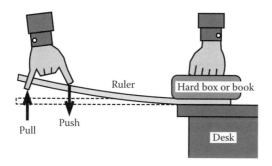

FIGURE 2.1.14 Experiment using a plastic ruler.

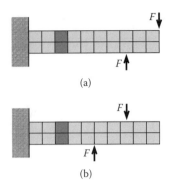

FIGURE 2.1.15 Cantilever with two loads in opposite directions.

EXERCISE USING GOYA-C:

Click the 'Details of Beam' button to obtain the window below specifying the length, width and height of the beam (Figure 2.1.16). Change the height to 7.5 mm, click the 'update' button, and click the main window. You will obtain Figure 2.1.17. Note that the tensile and compressive forces on the yellow segment double from 50 to 100 N because the distance between the forces ($(2/3)h$ in Figure 2.1.11) is halved from 10 to 5 mm, while the bending moment $M = T.(2/3)h$ remains the same. The increase of the forces causes larger flexural deformation (8 times). This phenomenon will be discussed in Section 2.8.

Design your own beam (Part 1)

We want to design a beam complying with the following conditions.

1. The magnitude of the external forces (F in Figure 2.1.18) shall be any single digit you choose plus 10 lbf.

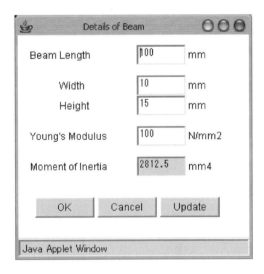

FIGURE 2.1.16 Window specifying the details of the beam.

Moments and Deflections in Cantilever Beams

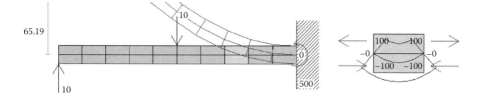

FIGURE 2.1.17 Beam with smaller height.

2. The distance between the forces shall be any single digit you choose plus 30 in.
3. The tensile force in the beam (*T* in Figure 2.1.18) shall not exceed 50 lbf.

What is the required beam height? Check your result using GOYA-C. You will understand that bending moment is an action that bends a beam (Figure 2.1.19).

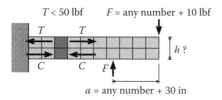

FIGURE 2.1.18 Your beam.

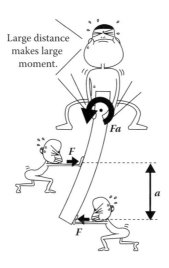

FIGURE 2.1.19 Bending moment is an action that bends a beam.

Bending moment and deformation

 Hi, Master. I still do not understand the action of a bending moment. I know that a tensile axial force elongates each atom in a member. Does a bending moment bend each atom?

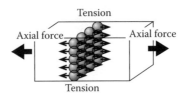

 No, it does not. Each atom in a bent member has a pair of tensile or compressive forces so that it elongates or shortens. The uppermost atom has the largest tensile forces, while the lowermost atom has the largest compressive force.

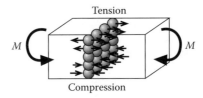

 Hmm, if the distances between atoms elongate or shorten, the beam must be a trapezoid, and will not bend, I believe.

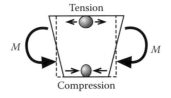

 Not true. Each atom is so small that you should slice the beam into very thin slices to understand what is happening. Each slice will be a trapezoid, and the beam will bend. Bend your eraser with your fingers. You will observe flexural deformation.

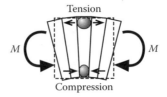

Moments and Deflections in Cantilever Beams

By the way, I don't like your Figure. Atoms do not deform in reality. The distance between atoms elongates or shortens.

Don't worry. I like it.

2.2 WHAT IS A SHEAR FORCE?

In section 2.1, we studied a case with two loads applied to a beam. In this section, we shall study a case with one load applied as shown in Figure 2.2.1a. Turn on GOYA-C and click 'Fade' to expose the forces within the beam (Figure 2.2.1b). As well as the bending moment of 800 N-mm, you will find a force of 10 N acting down. This force balances the external load of 10 N at the end of the beam. This type of internal force acting on the section is called a *shear force** and is usually denoted by the letter V.

Note that the distance between the external load and the section we have cut is 80 mm (8 segments). The external load and the shear force cause a clockwise moment of $10 \times 80 = 800$ N-mm. This moment, acting on the beam, is balanced by the internal (resisting) bending moment of equal magnitude but opposite sense at the cut.

---

* Similar to axial force or bending moment, you can represent and "see" a shear force only if you "cut" the beam and consider equilibrium at the section of the cut. While the part of the beam to the left of the cut is subjected to a downward shear force and counterclockwise bending moment (Figure 2.2.1b), the part of the beam to the right of the cut is subjected to an upward shear force and clockwise bending moment as shown in Sketch (a) below. Shear force is not a vector but a set of two forces shown in Sketch (b) below. Its effect is similar to that of a pair of shears (or scissors). We shall revisit this topic at the end of this section.

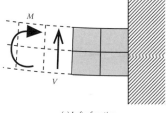

(a) Left of section

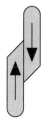

(b) Shear

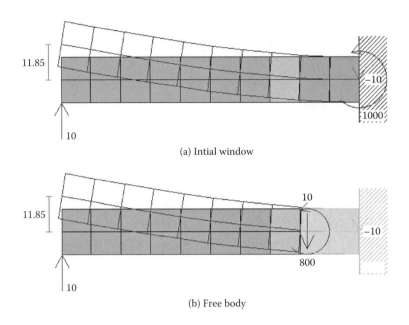

FIGURE 2.2.1 Beam with load at the free end.

Move your cursor to any grid and click. The shear force remains at 10 N, while the bending moment varies. For example, at a distance of 40 mm from the beam end, the moment acting on the section is 400 N-mm (Figure 2.2.2).

Let the distance from the free end to the section be x (Figure 2.2.3a). Equilibrium of moments of the free body shown in Figure 2.2.3b leads to

$$M = V.x \quad (2.2.1)$$

Bending moment M is distributed as shown in Figure 2.2.3c: it is zero at the free end ($M = 0$ at $x = 0$) and increases linearly to the fixed end ($M = V.L$ at $x = L$). The symbol $\cup$ in the diagram is used to indicate that the bending moment compresses the top fiber of the beam.

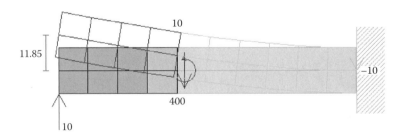

FIGURE 2.2.2 Shear force and bending moment at a section.

Moments and Deflections in Cantilever Beams

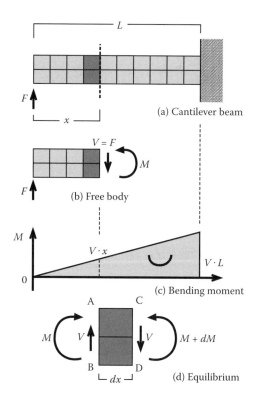

FIGURE 2.2.3 Shear force and bending moment in a cantilever determined using face-body diagrams.

Differentiating Equation 2.2.1 with respect to x, we obtain

$$dM/dx = V \qquad (2.2.2)$$

This relationship is very important because it is valid for any beam or column subjected to any set of loads. Equation 2.2.2 states that the rate of change of moment with distance is the shear force. Conversely, the integration of shear over a distance results in moment.

We can also derive Equation 2.2.2 considering equilibrium of a short length (length $= dx$) shown in Figure 2.2.3d which corresponds to the highlighted (yellow in GOYA) segment in Figure 2.2.3b. Equilibrium of the forces in the vertical direction requires the existence of a downward shear force V in section CD (Figure 2.2.3d). Call the bending moment in section CD ($M + dM$) and consider the equilibrium of moments as follows.

Bending moment at section AB: M, clockwise;*
Bending moment at section CD: $M + dM$, counterclockwise; and

---

* Bending moment is an action that bends beam and does not have a sense (direction). However, if you cut the beam and look at the section at the cut, the bending moment has a sense (clockwise or counterclockwise). Recall that an internal axial force also has a sense only when you cut the member and consider equilibrium at the section of the cut.

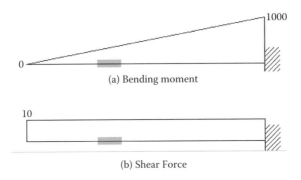

(a) Bending moment

(b) Shear Force

FIGURE 2.2.4 Bending moment and shear force diagram.

Moment caused by the shear forces in sections AB and CD: $V.dx$, clockwise. Defining clockwise moments to be positive, we have the following equation:

$$M - (M + dM) + V.dx = 0 \quad (2.2.3)$$

which again leads to $dM/dx = V$.

In the lower part of the window, you can find the distribution of the bending moment (linear) and the shear force (constant). Because the length of the beam is 100 mm, the slope of the line denoting the variation of moment along the span of the beam is $1000/100 = 10$ N and is equal to the shear force.

Click at various positions along the beam to see the flexural deformation of each grid as shown in Figure 2.2.5. You will find that the flexural deformation is larger in segments closer to the wall, because the bending moment is larger there. If you apply an upward force at the end of a wood beam as shown in Figure 2.2.6, it will break at the fixed end because the bending moment is largest there. If you increase the length L, the breaking force required will be smaller because, for the same force, the bending moment at the fixed end will be larger.

This phenomenon is equivalent to the action of a crow-bar (Figure 2.2.7). If you increase L, the force pulling the nail T will increase. Note that the sense or direction of action of the tensile force T is shown as being to the right because the force represents the resistance of the nail. The wall pushes against the heel of the crow-bar

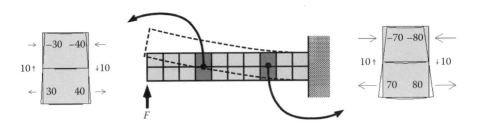

FIGURE 2.2.5 Grid deformations.

Moments and Deflections in Cantilever Beams

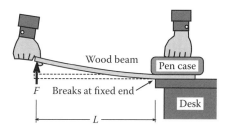

FIGURE 2.2.6 Breaking a wood beam.

with a force having a magnitude of C (= T). The wall also applies a downward force equal to F on the crow-bar.

Move the load to the middle of the beam as shown in Figure 2.2.8. You will notice that the bending moment becomes zero to the left of the load.

Figure 2.2.9 provides the reason. If you cut the beam to the left of the force as shown in Figure 2.2.9a, you will find that the part to the left of the cut is free of any force: it moves upward but does not deform. If you cut the beam at a distance of x to the right of the force as shown in Figure 2.2.9b, you will find that the section resists a shear force of $V = F$ and a bending moment of $M = V.x$.

Click 'Add load'. As shown in Figure 2.2.10, you will find that the shear force distribution (bottom right of the window) changes abruptly at the loading point. You will also find that the slope of the bending moment changes at the loading point (bottom left of the window).

We can explain this change by superposing the effect of F_1 on that of F_2 (Figures 2.2.11a and b). Note that the slope of the bending moment between points A and B is $dM/dx = 500/50 = 10$ N and the slope between points B and C is $dM/dx = (1500–500)/50 = 20$ N. Each slope agrees with the shear force in each region.

Click load F_2 and change it to -10 N as shown in Figure 2.2.12 using the bar in the bottom right. This is the loading we studied in Section 2.1. Note that the shear is zero in the right half of the beam. Also note that the bending moment is constant in

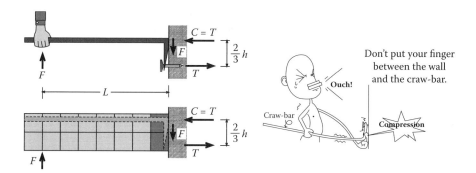

FIGURE 2.2.7 Equilibrium of crow-bar.

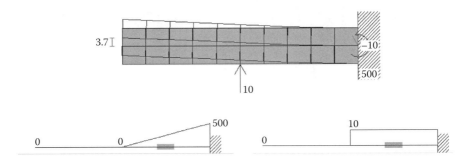

FIGURE 2.2.8 Move the load to the middle.

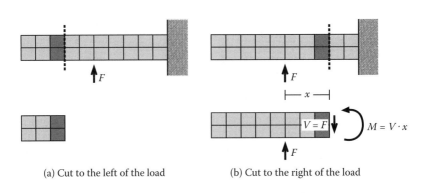

(a) Cut to the left of the load (b) Cut to the right of the load

FIGURE 2.2.9 The beam cut at two sections.

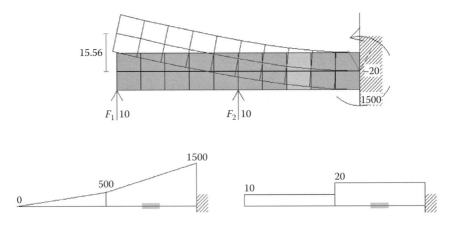

FIGURE 2.2.10 Shear force and bending moments in a cantilever with two upward loads.

Moments and Deflections in Cantilever Beams

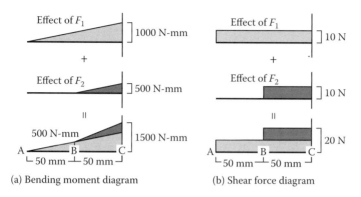

FIGURE 2.2.11 Superposition.

the right half of the beam because $dM/dx = V = 0$. Click at various points along the right half of the beam and find that each segment deforms the same amount because the bending moment is constant.

The bending moment and the shear force diagrams in Figure 2.2.12 can again be explained by the superposition shown in Figure 2.2.13. Note that the slope of the bending moment between points B and C is $dM/dx = (500-500)/50 = 0$ N, which agrees with the shear force in the region.

Look at the truss in Figure 2.2.14. Between points A and B, the top and bottom chords resist axial forces of -50 N (compression) and $+50$ N (tension), respectively. This corresponds to uniform bending moment. Between points A and B, the inclined members do not resist any axial load because the shear force is zero. The axial forces in the top and bottom chords decrease from a maximum at B to a minimum at C, while the axial forces in the inclined members remain constant a $10\sqrt{2}$ N, balancing to the uniform shear force of 10 N.

Next, go back to the beam and change force F_2 to -20 N as shown in Figure 2.2.15. The bending moment at the wall becomes zero, and the shear force in the right half of the beam becomes -10 N (negative). Recall the equation, $dM/dx = V$, again. The bending moment decreases at the rate of -10 N in the right half of the beam.

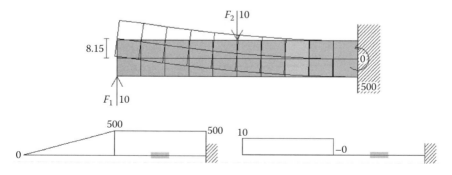

FIGURE 2.2.12 Cantilever with loads of equal magnitude and opposite direction.

96   Understanding Structures: An Introduction to Structural Analysis

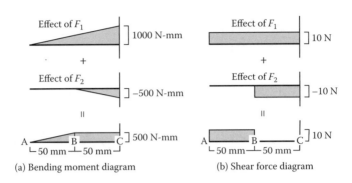

(a) Bending moment diagram   (b) Shear force diagram

FIGURE 2.2.13 Superposition.

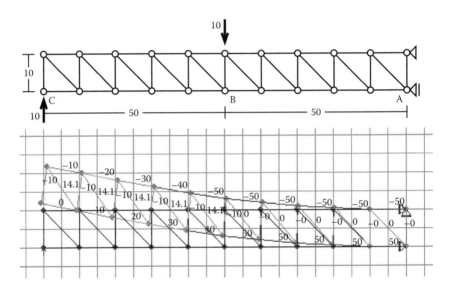

FIGURE 2.2.14 Cantilever truss.

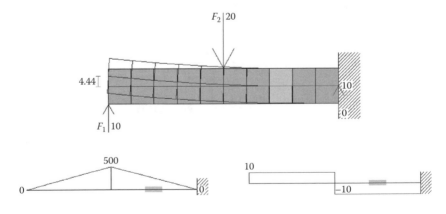

FIGURE 2.2.15 Change the load at mid-span to −20 N.

Moments and Deflections in Cantilever Beams

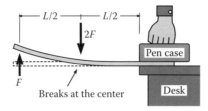

FIGURE 2.2.16 Break a wood beam using two opposing forces with the one at mid-span twice the one at the end.

If you apply a set of forces to a wood beam as shown in Figure 2.2.16, the bending moment has the largest value at mid-span so that the beam will break at the middle as a result of the fracture of the bottom fiber. In general, we can predict where a beam will break using the bending moment diagram.

In Chapter 1, we defined the axial force to be positive if tensile and negative if compressive. Here, we shall define the shear force to be positive if it acts clockwise as shown in Table 2.1. Also, we shall define the bending moment to be positive if it compresses the top fiber of the beam.

Click and change the force F_2 to −40 N as shown in Figure 2.2.17. The shear force in the right half of the beam becomes −30 N. Because $dM/dx = -30$, the bending moment at the face of the wall becomes negative, indicating that the bottom fiber is in compression. You should remember that the position of the bending moment diagram with respect to the axis of zero moment indicates which face of the beam is compressed edge. If the moment diagram is below the line of zero moment, the bottom edge of the beam is in compression. If the moment diagram is above the line of zero moment, the top edge of the beam is in compression.

If you apply a set of forces on a wood beam as shown in Figure 2.2.18 so that the top force is four times as large as the bottom force, the bending moment will have the largest value at the fixed end and the beam will break at that point as a result of tensile failure of the top fiber.

TABLE 2.1
Notations

	Axial force	Shear force	Bending moment
+	Tension $N \leftarrow \square \rightarrow N$	Clockwise $V \uparrow \square \downarrow V$	Top compression
−	$N \rightarrow \square \leftarrow N$ Compression	$V \downarrow \square \uparrow V$ Counter-clockwise	Bottom compression

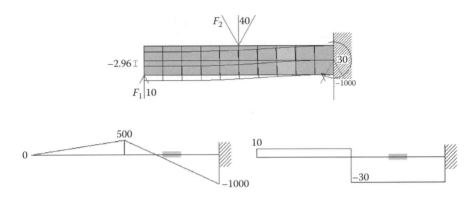

FIGURE 2.2.17 Change the load at mid-span to −40 N.

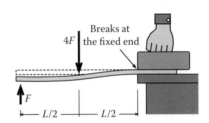

FIGURE 2.2.18 Break a wood beam using forces of F applied at the free end and $4F$ applied at mid-span.

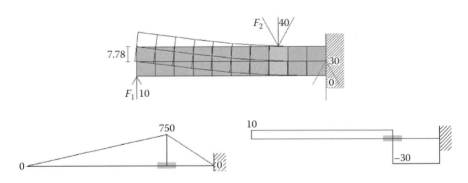

FIGURE 2.2.19 Move the −40 N load at mid-span to the right.

Moments and Deflections in Cantilever Beams

Click and move the force F_2 to the right as shown in Figure 2.2.19. When it reaches a point 25 mm (two and a half grids) from the fixed end, the bending moment at the wall becomes zero. Note that $dM/dx = 750/75 = 10$ and $dM/dx = -750/25 = -30$ to satisfy $dM/dx = V$ again.

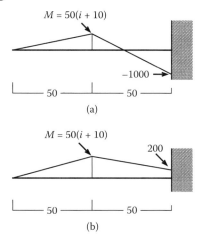

FIGURE 2.2.20 Bending moment diagrams.

EXERCISE 2.2.1 USING GOYA-C

Use any number i you may choose to define the bending-moment diagrams in Figure 2.2.20a,b. Find the loads that cause the bending-moment diagrams. And sketch the deformed shapes for the two cases.

EXERCISE 2.2.2 USING GOYA-C

If you select the locations and the magnitude of the forces carefully, you may create interesting bending moment diagrams as those illustrated in Figure 2.2.21. Make up your own diagram, and report the corresponding loads and the deformations. If the deformation of your beam becomes too large, you can adjust it decreasing the number in the 'Amplification' text-field in the bottom left.

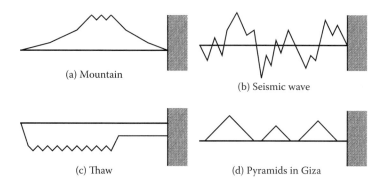

FIGURE 2.2.21 Interesting bending-moment diagrams.

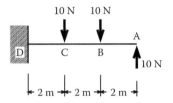

FIGURE 2.2.22 A cantilever with forces applied at A, B, and C.

EXAMPLE 2.2.1

Calculate the shear force and the bending moment diagrams for the loads shown in Figure 2.2.22, and sketch the deformation of the beam.

Solution

Cut the beam between the points A and B. From the conditions of equilibrium, you will find that the shear force is −10 N at any section between A and B (negative because it is counter-clockwise). Cuts between B-C and C-D yield the shear force diagram in Figure 2.2.23a. Noting that the bending moment at the free end is zero and the slope is $dM/dx = -10$ between A and B, we get $M = -10 \times (-2) = 20$ N-m at B. The bending moment between B and C should be constant because the shear force is zero from B to C. Using the same logic we get $M = 0$ at D.

The bending moment diagram in Figure 2.2.23b indicates that the top fiber of the beam is in compression between A and D. Noting that the fixed end of the beam

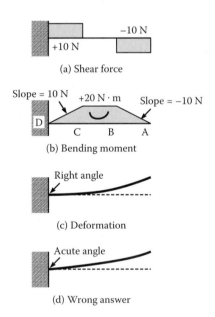

FIGURE 2.2.23 Shear and moment diagrams for the cantilever in Figure 2.2.22 and correct and incorrect solutions for the deflected shape.

Moments and Deflections in Cantilever Beams

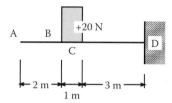

FIGURE 2.2.24 Shear-force diagram in a cantilever coused by two loads of unknown magnitude and direction.

is perpendicular to the wall, we can reason that the beam will deform as shown in Figure 2.2.23c. The deflected shape shown in Figure 2.2.23d is not correct because the deflected beam makes an acute angle with the wall.

EXAMPLE 2.2.2

Calculate the loads and the bending moment diagrams corresponding to the shear force diagram shown in Figure 2.2.24, and sketch the deflected shape of the beam.

Solution

Because the shear force changes at load points, there should be applied loads of 20 N at points B and C (and nowhere else). Noting that the shear force is positive (or clockwise), we conclude that the direction of the loads should be as shown in Figure 2.2.25a.

Because there is no load between A and B, the bending moment is zero from A to B (Figure 2.2.25b). It increas es at the rate of $dM/dx = 20$ between B and C to reach 20 N-m at C. Between C and D, it is constant because $V = 0$ from B to C.

Noting that the bending moment is positive between B and D, we reason that the beam will deform as shown in Figure 2.2.25c. Note that the base of the beam is

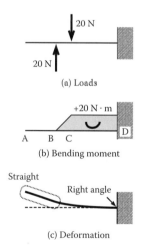

FIGURE 2.2.25 Applied loads, moment distribution, and deflected shape for the shear-force diagram shown in Figure 2.2.24.

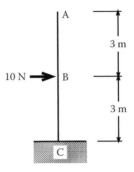

FIGURE 2.2.26 A vertical cantilever fixed at C and subjected to a horizontal load at its mid-height.

perpendicular to the wall. The beam does not deform (remains straight) between A and B because the bending moment is zero in that portion of the beam.

EXAMPLE 2.2.3

Calculate the shear force and the bending moment corresponding to the loads shown in Figure 2.2.26, and sketch the deflected shape of the beam. (Hint: Recall that the sense of the bending moment diagram gives us a clue as to which face—left or right— of the beam is compressed.)

Solution

The portion between A and B is free of shear force and bending moment. Between B and C, the load induces a clockwise shear force (Figure 2.2.27a). The beam bends between B and C. The deformed shape is shown in Figure 2.2.27c. The bending moment at C is 10 × 3 = 30 N·m. Noting that the sense of the bending-moment diagram indcates which side of the beam is in compression, we have Figure 2.2.27c. If we define the positive values of the coordinates of x and M as shown by the arrows in Figure 2.2.27c, the Figure satisfies the relation $dM/dx = V$.

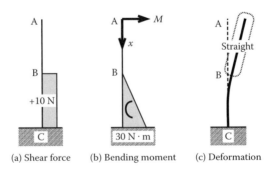

FIGURE 2.2.27 Shear-force and bending-moment diagrams and deflected shape for the cantilever in Figure 2.2.24.

Moments and Deflections in Cantilever Beams

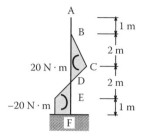

FIGURE 2.2.28 Bending-moment diagram in a vertical cantilever caused by unknown horizontal forces.

EXAMPLE 2.2.4

Calculate the loads and the shear force diagrams corresponding to the bending moment diagram shown in Figure 2.2.28, and sketch the deflected shape of the beam.

Solution

Defining the positive values of x and M as we did in Figure 2.2.27b, we have $V = dM/dx = 10$ in B–C, $V = dM/dx = -20$ in C–E, and $V = 0$ in E–F as shown in Figure 2.2.29a. Recalling that the positive shear force is clockwise, we have the loads as shown in Figure 2.2.29b. Because the right edge of the beam is compressed between B and D and the left edge between D and F, we reason the deflected shape to be as shown in Figure 2.2.29c.

Design your own beam (Part 2)

We want to design a beam that can support a mini-elephant. The weight of the mini-elephant is (any one digit you choose) plus 10 lbf. Assume that each leg of the mini-elephant carries the same amount of gravity force. The weight of the beam is negligible. The tensile force on the wall (T in Figure 2.2.30) must be smaller than

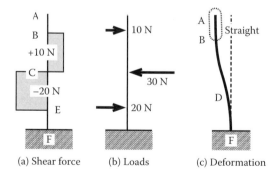

FIGURE 2.2.29 Shear-force diagram, applied loads, and deflected shape corresponding to the moment diagram in Figure 2.2.26.

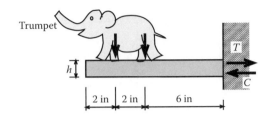

FIGURE 2.2.30 An elephant on a cantilever.

100 lbf because of the strength of the wall. What is the required beam depth, h? (Check your result using GOYA-C.)

What is an internal shear force?

I could not understand the definition of shear force in today's lecture. Is it an action that shears a beam as shown in Sketch (a)?

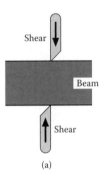

No, it is different. A pair of shears acts to cut the beam from outside, whereas the shear force is not an action from outside. It acts inside the beam as shown in Sketch (b). You cannot "see" the forces unless you cut the beam over a length dx. Similarly, the shear force refers to a force pair acting perpendicularly to the axis of a structural member as shown in Sketch (c). The force pair is internal. It is not applied externally.

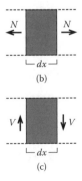

Moments and Deflections in Cantilever Beams

You mean that shear force is similar to axial force. But it looks different. In the case of tensile axial force, the pair of forces balance each other by pulling the portion dx in opposite directions. In the case of shear force, however, the two forces balance each other in the vertical direction and they create a moment V.dx that tends to rotate the beam portion dx.

You have got it. The product of V and dx results in an incremental bending moment, dM = V.dx that is, a shear force always leads to a slope in the bending-moment diagram dM/dx.

I know that an axial force elongates or shortens a member and a bending moment bends a member. How does a shear force deform a member?

Good question. Assume that you cut the beam and glue the sections using a very soft adhesive. Then, apply shear force. The adhesive will deform as shown in Sketch (e) because the atoms in the adhesive are subjected to the forces shown in Sketch (f). We call the deformation "shear deformation."

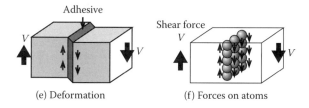

(e) Deformation   (f) Forces on atoms

Note that this is a special case occurring for very small ratios of moment to shear (M/V). As you can see in Sketch (g), tensile and compressive forces caused by bending moment are usually much larger than shear force. So, shear deformation is usually much smaller than flexural deformation. But shear force is important in determining the bending moment. Also, shear force sometimes induces a brittle failure. You should never ignore it.

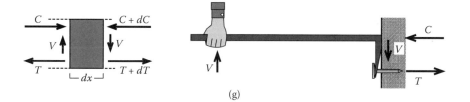

(g)

2.3 WHAT IS A DISTRIBUTED LOAD?

A cantilever beam may deflect under snow load in winter. In summer, strong winds may push the beam up. We call such loads 'distributed loads'. Log into GOYA-C. Type 0.5 in the text field for 'Amplification' to reduce the scale. Click 'Add load' nine times and move each load to develop the distributed load shown in Figure 2.3.1a. Note that the width of each element is 10 mm and the magnitude of each load is 10 N. This represents a distributed load of (10 N)/(10 mm) = 1 N/mm. The shear-force diagram (Figure 2.3.1c) is shown at intervals of 10 mm. You will have smoother diagrams if you take the time to specify 100 loads of 1 N at intervals of 1 mm.

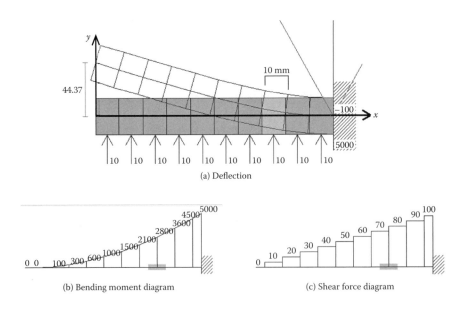

FIGURE 2.3.1 Simulation of distributed load.

Moments and Deflections in Cantilever Beams

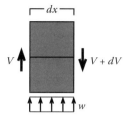

FIGURE 2.3.2 Equilibrium of *dx*.

Figure 2.3.2 shows the vertical forces acting on a segment of length dx where we assume that the distributed force w acts on every inch (or mm) along the beam.* Equilibrium in the vertical direction requires that $dV = w.dx$ or

$$dV/dx = w \qquad (2.3.1)$$

If w is constant (uniformly distributed load), integration of the above equation leads to $V = w.x + C$, where C is constant. Because the shear force is zero at the free end as shown in Figure 2.3.1c, the constant C should be zero and we have

$$V = w.x \qquad (2.3.2)$$

Substituting this into $dM/dx = V$ and noting that $M = 0$ at the free end ($x = 0$), we have

$$M = w.x^2/2 \qquad (2.3.3)$$

which describes a parabola. Note that the bending moment diagram in Figure 2.3.1 is almost parabolic. If we substitute $w = 1$ N/mm and $x = 100$ mm (the beam length), we have $V = 100$ N and $M = 5000$ N-mm, which agree with the numbers indicated in Figure 2.3.1c,b at the fixed end.

We may also obtain these equations without integration. Cut the beam as shown in Figures 2.3.3a and b, and represent the distributed load by an equivalent concentrated force as shown in Figure 2.3.3c. Then, equilibrium in the vertical direction leads to Equation 2.3.2 and moment equilibrium leads to Equation 2.3.3.

We can approximate the distributed force by a few concentrated forces as shown in Figure 2.3.3d. The broken and solid lines in Figure 2.3.3e show the exact and approximated shear force diagrams, respectively. They agree at the free end, midspan, and at the fixed end. The bending moment diagrams also agree at the same points as shown in Figure 2.3.3f. Simulate the concentrated forces shown in Figure 2.3.3d on GOYA-C. You will find that the deflections are also similar to those shown in Figure 2.3.1a.

---

* In Figure 2.3.2, the symbol ⭡⭡⭡⭡⭡ is meant to indicate that the forces are distributed uniformly.

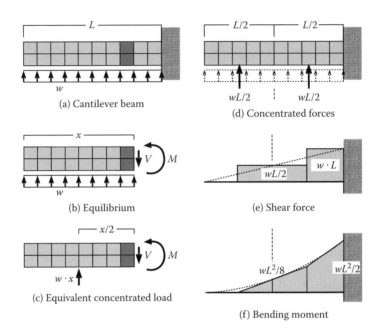

FIGURE 2.3.3 Beam with uniformly distributed load.

EXAMPLE 2.3.1

Assume that you have a beam with the same magnitude of distributed load as the beam in Figure 2.3.3a but double the span L. Determine the shear force and the bending moment at the fixed end.

Solution

According to Equations 2.3.2 and 2.3.3, the shear force will double and the bending moment will be four times that in the previous example.

EXAMPLE 2.3.2

Draw shear-force and bending-moment diagrams caused by the load shown in Figure 2.3.4.

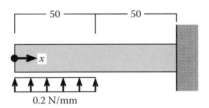

FIGURE 2.3.4 Uniform load distributed over half of cantilever span.

Moments and Deflections in Cantilever Beams

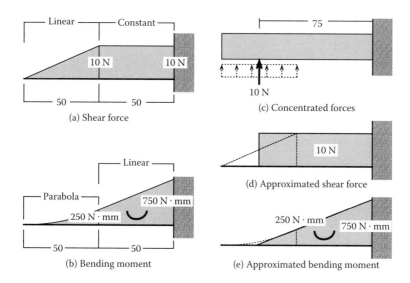

FIGURE 2.3.5 Solution.

Solution

We have $dV/dx = 0.2$ in the left half ($0 \leq x \leq 50$) and $V = 0$ at $x = 0$. So,

$$V = 0.2\,x \qquad \text{in } 0 \leq x \leq 50 \tag{2.3.4}$$

as shown in Figure 2.3.5a. For $50 \leq x \leq 100$, $dV/dx = 0$ yields $V =$ constant. Substituting Equation 2.3.4 into $dM/dx = V$ and noting $M = 0$ at the free end ($x = 0$)

$$M = 0.1\,x^2 \qquad \text{in } 0 \leq x \leq 50.$$

as shown in Figure 2.3.5b. Because $V = 10$ in $50 \leq x \leq 100$, the slope of the bending-moment diagram should be 10. Figure 2.3.5c shows the concentrated force equivalent to the distributed load. Figures 2.3.5d and e show the approximated shear force and bending moment diagrams. Note again that they agree with the exact values at the free end, mid-span, and at the fixed end.

EXAMPLE 2.3.3

Draw shear-force and bending-moment diagrams caused by the distributed force in Figure 2.3.6. Note that w is negative because the force is downward.

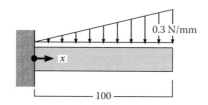

FIGURE 2.3.6 Linearly varying distributed load.

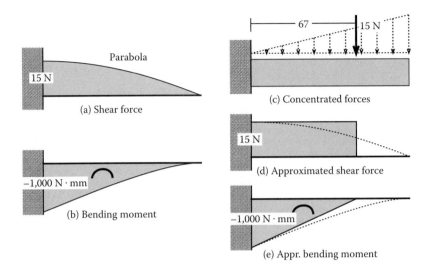

FIGURE 2.3.7 Solution.

Answer

Because the load is $w = -(0.3/100)x$, we integrate $dV/dx = -(0.3/100)x$ noting $V = 0$ at the free end ($x = 100$) and obtain

$$V = -15x^2 \times 10^{-4} + 15$$

Substituting this into $dM/dx = V$ and noting $M = 0$ at $x = 100$

$$M = -5x^3 \times 10^{-4} + 15x - 1000$$

These results are shown in Figures 2.3.7a and b. Figure 2.3.7c shows the concentrated force equivalent to the distributed load. The force has the magnitude equal to the triangle in Figure 2.3.6,

$$F = \frac{1}{2} \times (0.3 \text{ N/mm}) \times (100 \text{ mm}) = 15 \text{ N}$$

and is located at the centroid of the triangle. Figures 2.3.7d and e show the approximated shear-force and bending-moment diagrams. They agree with the exact values at the free end and at the fixed end.

EXAMPLE 2.3.4

Draw shear-force and bending-moment diagrams caused by the loads shown in Figure 2.3.8. (Hint: you may draw the diagrams for the distributed and concentrated loads separately and add them.)

Moments and Deflections in Cantilever Beams

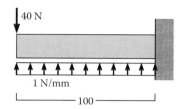

FIGURE 2.3.8 Distributed load plus concentrated load.

Solution

As we noted in Section 2.2, the shear force and bending-moment diagrams caused by the concentrated load (−40 N) are plotted as shown in Figure 2.3.9. Those related to the distributed load are shown in Figure 2.3.10. The areas in the shear-force diagrams equal the bending moments ($\int V \cdot dx = M$). Adding (superposing) these diagrams, we obtain Figures 2.3.11a and b. Note that the shear force and the slope of the moment diagram are zero at $x = 40$ mm. Figure 2.3.11c shows the concentrated forces equivalent to the distributed load. Figures 2.3.11d and e show the approximate shear-force and bending-moment diagrams. They agree with the exact values at the free end, mid-span, and at the fixed end. Figure 2.3.12 shows the results obtained by GOYA-C. Note that the deflection at the beam end is only 3 mm, which is less than 1/10 of the deflection caused by the distributed load shown in Figure 2.3.1 (44 mm). The clue to explain this result may be found in the bending-moment diagram, which shows that the beam deflects up near the fixed end and down near the middle. The upward deflection tends to reduce the downward deflection.

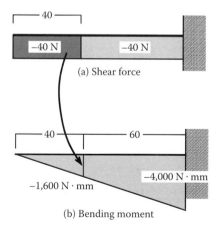

FIGURE 2.3.9 Contribution of concentrated load.

112 Understanding Structures: An Introduction to Structural Analysis

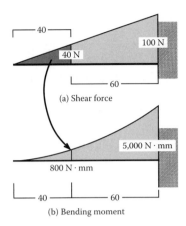

FIGURE 2.3.10 Contribution of uniformly distributed load.

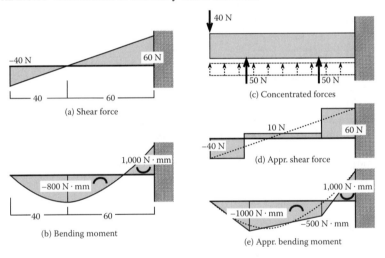

FIGURE 2.3.11 Solution.

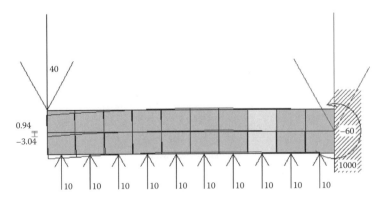

FIGURE 2.3.12 Simulation using GOYA-C.

Moments and Deflections in Cantilever Beams

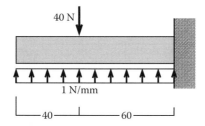

FIGURE 2.3.13 Cantilever with distributed and concentrated loads.

EXAMPLE 2.3.5

Plot shear-force and bending-moment diagrams caused by the load in Figure 2.3.13. (Do not use GOYA-C before making your own calculations.)

Solution

The shear-force and bending-moment diagrams caused by the concentrated load (–40 N) are given as shown in Figure 2.3.14. Adding this diagram to the one in Figure 2.3.10, we obtain Figure 2.3.15. Note that the shear force is discontinuous at $x = 40$ mm. Figure 2.3.16 shows the results simulated by GOYA-C. Note that the deflection at the beam end is 24 mm, which is eight times that obtained for the loading condition in Figure 2.3.12 (3 mm). The reason lies again in the bending moment diagram, which shows that the beam deflects upward throughout its length. Approximate the bending moment diagram using concentrated forces equivalent to the distributed load and compare with the exact diagram.

We have distinguished the effects of a concentrated load from those of a distributed load. However, if we look at the experiment shown in Figure 2.3.17 closely, we may notice that the finger applying the load has a finite length in the direction of the span. The applied load should be modeled as a distributed load as shown in Figure 2.3.18a. Such modeling leads to the shear-force diagram shown in Figure 2.3.18b. In other words, the equation $dV/dx = w$ also applies to concentrated loads.

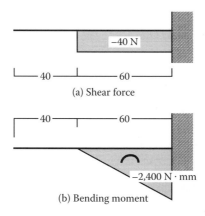

FIGURE 2.3.14 Contribution of concentrated load.

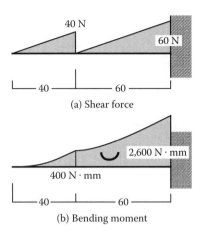

FIGURE 2.3.15 Solution.

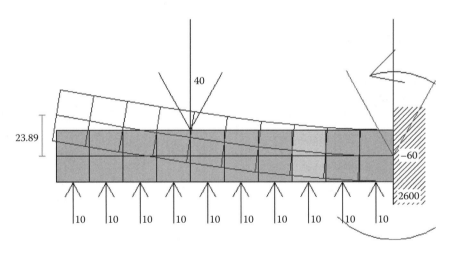

FIGURE 2.3.16 Simulation using GOYA-C.

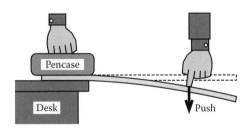

FIGURE 2.3.17 Push a cantilever beam.

Moments and Deflections in Cantilever Beams

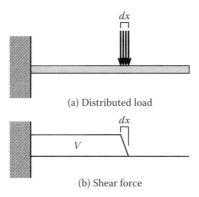

FIGURE 2.3.18 Distributed load.

EXAMPLE 2.3.6

Draw shear-force and bending-moment diagrams caused by the distributed load of varying magnitude shown in Figure 2.3.19.

Solution

Integrating the distributed load leads to the shear force.

$$V = \int w \cdot dx = \frac{20}{\pi}\cos\left(\frac{\pi}{2}x\right) + C_1$$

where C_1 is an integral constant. Because $V = 0$ at the free end ($x = 0$), we have

$$C_1 = -\frac{20}{\pi}$$

Figure 2.3.20a shows the shear force diagram. Integrating the shear force leads to the bending moment.

$$M = \int V \cdot dx = \frac{40}{\pi^2}\sin\left(\frac{\pi}{2}x\right) - \frac{20}{\pi}x + C_2$$

where C_2 is an integral constant. Because $M = 0$ at the free end ($x = 0$), we have $C_2 = 0$. Figure 2.3.20b shows the bending-moment diagram. Because the diagram

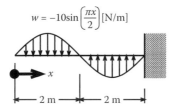

FIGURE 2.3.19 Cantilever with distributed load of varying magnitude.

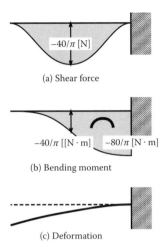

FIGURE 2.3.20 Exact solution.

indicates that the bottom of the beam is compressed, the beam deflects downward as shown in Figure 2.3.20c.

Let us represent the distributed load by a pair of concentrated loads as shown in Figure 2.3.21a, noting that

$$\int_0^2 \sin\left(\frac{\pi}{2}x\right)dx = \frac{40}{\pi}$$

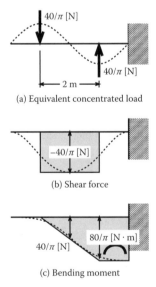

FIGURE 2.3.21 Approximate solution.

Moments and Deflections in Cantilever Beams

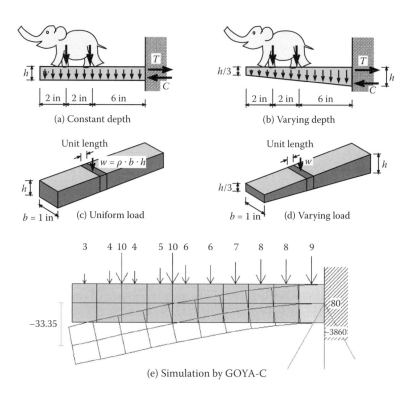

FIGURE 2.3.22 Design your beam.

The shear-force and bending-moment diagrams for these loads are shown in Figure 2.3.21b and c. They agree with the exact values at the free end, at mid-span and at the fixed end.

Design your own beam (Part 3)

We want to design a beam that can carry a mini elephant, whose weight is digit you select plus 10 lbf. Assume that each leg carries the same amount of gravity force. The density of the beam is 0.5 lbf/in³. The tensile force on the wall (T in Figure 2.3.22a) must be smaller than 90 lbf. What is the required beam depth if the beam is prismatic (its section remains the same along its span) (Figure 2.3.22a) or if it has varying depth as specified in Figure 2.3.22b? Check your results using GOYA-C.

Hint: If the depth is constant, the distributed load caused by the self-weight is $w = \rho \cdot b \cdot h$,

where

ρ: density (0.5 lbf/in.³), b: beam width (10 in.), and

h: beam depth (unknown)

as shown in Figure 2.3.22c. The load w will vary if the beam depth varies. In both cases, you will derive equations in terms of h and solve them. The depth required for case b (varying) is smaller than that required for case a (uniform). You can find beams with varying depths in actual structures such as those supporting balconies or elevated highways. GOYA-C cannot simulate a beam with varying depth. You can check your result by using a varying load as shown in Figure 2.3.22e with the depth at the fixed end, h.

2.4 WHAT IS A COUPLE?

Apply a pair of equal and opposite forces of 100 N close to one another as shown in Figure 2.4.1. Note that a shear force exists only between the points of application of the forces. The bending moment increases at a high rate in that region. We call the pair a '*couple*'. If the distance (a in Figure 2.4.2a) is small enough, we represent the couple using a curved arrow as shown in Figure 2.4.2b. A couple may be applied through a bolt embedded in the side of the beam and twisted by a wrench. The bending moment has a sudden change or a "jump" at the point of application of the couple as shown in Figure 2.4.2c. We ignore the shear force associated with a couple though in fact a very large shear force might exist over a very small region (see the shear-force diagram in Figure 2.4.1).

Figure 2.4.3 shows a beam subjected to a couple and a concentrated load. The shear-force diagram is constructed ignoring the couple because it is assumed that the couple is applied at a point. The bending moment diagram has a jump at the point of application of the couple. The bending-moment diagram in Figure 2.4.3c is obtained combining the moment diagram in Figure 2.4.2c with that for load F (the broken line in Figure 2.4.3c). Figure 2.4.4 shows the simulation by GOYA-C for $M = 500$ N-mm and $F = 10$ N. If you use forces of ± 1000 N with a distance of 1 mm apart, you will have a moment diagram very similar to Figure 2.4.3c.

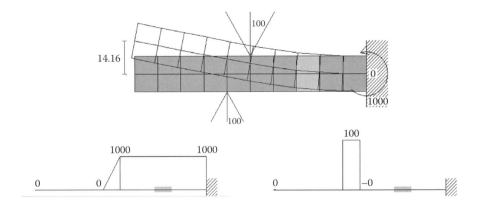

FIGURE 2.4.1 A pair of opposing loads placed close to one another.

Moments and Deflections in Cantilever Beams

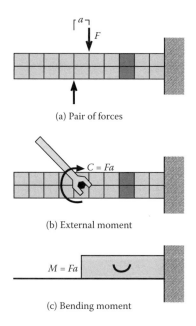

FIGURE 2.4.2 Beam with a couple.

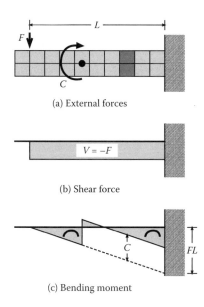

FIGURE 2.4.3 Beam with a couple and a load.

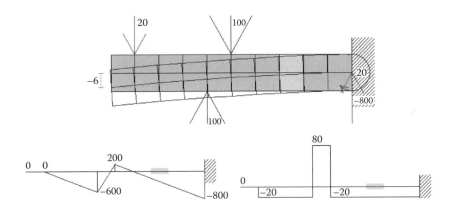

FIGURE 2.4.4 Simulation by GOYA-C.

EXAMPLE 2.4.1

Construct the shear-force diagrams and evaluate the loads corresponding to the bending moment in Figure 2.4.5, where

$$M = 20\cos\left(\frac{\pi}{2}x\right) \quad [\text{N.m}]$$

Solution

Differentiate the expression for bending moment to obtain the shear force at any point x.

$$V = \frac{dM}{dx} = -10\pi \sin\left(\frac{\pi}{2}x\right)$$

The resulting expression is plotted in Figure 2.4.6a. Differentiate the expression for the shear force to obtain the distributed load.

$$w = \frac{dV}{dx} = -5\pi^2 \cos\left(\frac{\pi}{2}x\right)$$

Because the bending moment at the free end is 20 N·m, there should be a couple at the end, and the load should be as shown in Figure 2.4.6b. The bending-moment diagram indicates that the beam deflects as shown in Figure 2.4.6c.

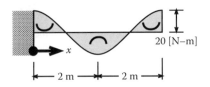

FIGURE 2.4.5 Bending-moment diagram.

Moments and Deflections in Cantilever Beams

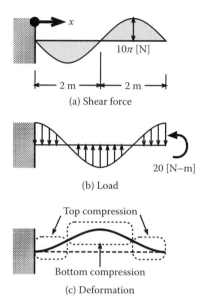

(a) Shear force

(b) Load

(c) Deformation

FIGURE 2.4.6 Solution.

The most important concepts in the Sections 2.1 through 2.4 are:

1. Shear force at a section is determined by "cutting" the beam at that section and considering force equilibrium in the direction perpendicular to the beam axis.
2. Bending moment at a section is determined by cutting the beam at that section and considering the moment equilibrium.

It is useful to remember that:

1. The slope of the bending-moment diagram along a member is equal to the shear force ($dM/dx = V$). A couple applied on the beam at a point makes the bending-moment diagram discontinuous (its magnitude changes abruptly) at that point.
2. The slope of the shear-force diagram along a member is equal to the distributed load ($dV/dx = w$). A concentrated force applied on the beam at a point makes the shear-force diagram discontinuous (a "jump" occurs) at that point.

Review: A couple is to bending moment as a force is to shear force.

In Section 1.2, we concluded that external force, a vector, is different from internal axial force, a scalar. If you apply forces parallel to the member axis (Figure 2.4.7a), the axial force changes where the forces are applied (Figure 2.4.7b). In the preceding section, we concluded that external force is different from shear force. If you

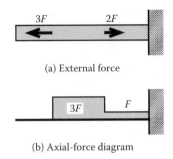

FIGURE 2.4.7 Axial force.

impose forces perpendicular to the member axis as shown in Figure 2.4.8a, the shear force changes where the external forces are applied (Figure 2.4.8b). Similarly, bending moment changes abruptly where a couple is applied as shown in Figure 2.4.9. A couple is a vector having a sense or direction with respect to the axis of rotation (Figure 2.4.10), while bending moment is a scalar.

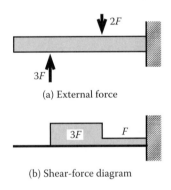

FIGURE 2.4.8 Shear force.

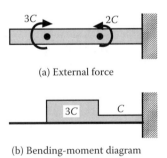

FIGURE 2.4.9 Bending moment.

Moments and Deflections in Cantilever Beams

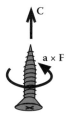

FIGURE 2.4.10 A couple is a vector.

2.5 THE EFFECTS OF MOMENT ON STRESSES IN A BEAM

In this section, we shall focus on bending moment based on the following definitions.

$$(\text{Force}) = (\text{Stress}) \times (\text{Area}) \tag{2.5.1}$$

$$(\text{Moment}) = (\text{Force}) \times (\text{Distance}) \tag{2.5.2}$$

Using GOYA-C, apply forces as shown in Figure 2.5.1 to have a bending moment of 600 N-mm at the fixed end. Click 'Stress' in the middle right to obtain Figure 2.5.2a. In Figure 2.5.2a we see that stress is distributed linearly over the depth of the section.* Figure 2.5.2b shows the isometric view of the stress distribution. The width, b, and the depth, h, of the beam section are 10 mm and 15 mm, respectively. The stress at the extreme fiber in compression (top fiber) is -1.6 N/mm² and the stress at the extreme fiber in tension (bottom fiber) is $+1.6$ N/mm². The objective of this section is to study how the assumed stress distribution on the section develops a bending moment of 600 N-mm.

The bending moment generated by the stress distribution can be determined if we partition the section into many layers of thin horizontal slices (Figure 2.5.3a). Because the thickness of each slice (dy) is small enough, we may assume that the stress in each slice is uniform. Consider the shaded slice in Figure 2.5.3a. It has an area, $b.dy$. Assuming a constant stress acting on the slice, we determine the total force on the slice as the product of the stress, σ, and the area of the slice $b.dy$: $dF = \sigma.b.dy$. The contribution of the force $dF = \sigma.b.dy$ to the bending moment is $dM = -y.dF = -\sigma.y.b.dy$ with a negative sign because the stress is compressive ($\sigma < 0$) in the upper half of the section ($y > 0$). The bending moment of the section is the sum of dM over the section or

$$M = -\int y \cdot dF = -\int_{-h/2}^{h/2} \sigma \cdot y \cdot b \cdot dy \tag{2.5.3}$$

Calling the stress at the bottom of the section σ_f and $-\sigma_f$ at the top (Figure 2.5.3b), we describe the stress distribution as follows.

$$\sigma = -\frac{y}{h/2} \times \sigma_f \tag{2.5.4}$$

---

* Strictly, the stress distribution is linear if the beam material remains in linear range where unit stress is proportional to unit strain. In this text we shall consider response only in the linear range.

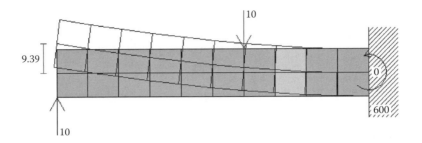

FIGURE 2.5.1 Forces making 600 N-mm.

Substituting this into Equation 2.5.3 leads to the resisting moment in terms of the stress in the extreme fiber, σ_f.

$$M = \int_{-h/2}^{h/2} \frac{y}{h/2} \cdot \sigma_f \cdot y \cdot b \cdot dy = \frac{2b\sigma_f}{h} \cdot \int_{-h/2}^{h/2} y^2 \cdot dy = \frac{bh^2}{6} \cdot \sigma_f \quad (2.5.5)$$

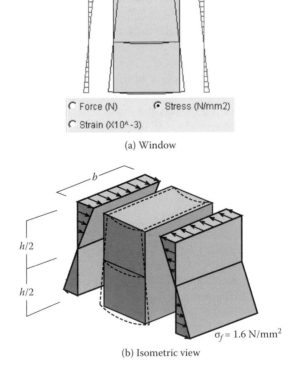

FIGURE 2.5.2 Stress distribution.

Moments and Deflections in Cantilever Beams

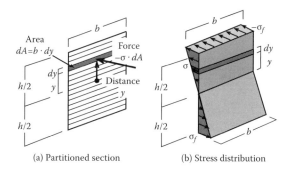

(a) Partitioned section   (b) Stress distribution

FIGURE 2.5.3 Partitioned section and stress distribution.

The term $bh^2/6$ modifying σ_f is called the "*section modulus*" of a rectangular section. We shall denote it by the symbol Z.

$$Z = \frac{bh^2}{6} \tag{2.5.6}$$

For a given moment and section, we express the maximum flexural stress in a rectangular section simply as:

$$\sigma_f = \frac{M}{Z} \tag{2.5.7}$$

The initial setting of GOYA-C is $b = 10$ mm and $h = 15$ mm, which leads to the following maximum bending stress for Figure 2.5.2.

$$\sigma_f = \frac{M}{Z} = \frac{600}{10 \times 15^2 / 6} = 1.6 \text{ N/mm}^2$$

Click 'Force' to get Figure 2.5.4a. The compressive force shown here represents the total stress in the upper half of the section (Figure 2.5.4b),

$$C = \int_0^{h/2} \sigma \cdot dA = \int_0^{h/2} \sigma \cdot b \cdot dy = \int_0^{h/2} \frac{y}{h/2} \cdot \sigma \cdot b \cdot dy = \frac{bh}{4} \cdot \sigma_f$$

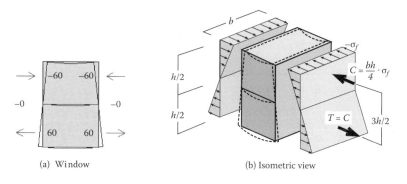

(a) Window   (b) Isometric view

FIGURE 2.5.4 Concentrated force.

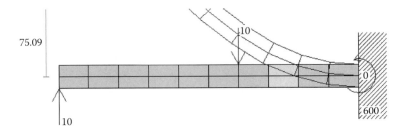

FIGURE 2.5.5 Beam depth is reduced to 7.5 mm.

The force is located at the centroid of the compressive stress block that is $h/3$ from the mid-height of the section. The tensile force is also located $h/3$ from the mid-height. The distance between the compressive and tensile forces is $2h/3$, which leads to

$$M = C \times \frac{2h}{3} = \frac{bh}{4}\sigma_f \times \frac{2h}{3} = \frac{bh^2}{6}\sigma_f = Z\sigma_f$$

Next, click on the 'Beam Detail' button and change the beam depth to 7.5 mm to get Figures 2.5.5 and 2.5.6. Figure 2.5.6 shows that the stress is 6.4 N/mm², which is 4 times that for $h = 15$ mm. Note that the stress is inversely proportional to the second power of the beam depth as indicated in Equation 2.5.6 and 2.5.7. Generally, the depth of a beam is more than its width because an increase in depth reduces the stress, for a given moment, more effectively than an increase in width.

EXAMPLE 2.5.1

A cantilever beam of width $b = 1$ in, depth $h = 1.5$ in, and length $L = 15$ in. is subjected to a force as shown in Figure 2.5.7a. Note that the selected dimensions correspond roughly to those of the bone in your forearm. Assume the material of the beam to be linearly elastic. Calculate the maximum load that can be resisted for the four cases of tensile strength listed below. Compare it in each case with the axial force that the beam can carry if it is loaded in axial tension as shown in Figure 2.5.7b.

(a) $\sigma = 15{,}000$ psi (human bone) (b) $\sigma = 10{,}000$ psi (pine)
(c) $\sigma = 60{,}000$ psi (steel) (d) $\sigma = 400$ psi (concrete)

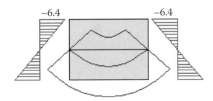

FIGURE 2.5.6 Stress distribution.

Moments and Deflections in Cantilever Beams

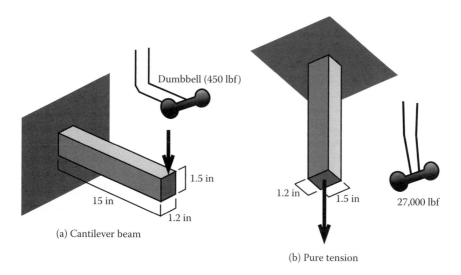

FIGURE 2.5.7 Cantilever subjected at its free end to (a) a transverse load and (b) an axial load.

Solution

The section modulus is $Z = bh^2/6 = 0.45$ in^2. Substituting the given material strengths into the equation $M = Z\sigma$, we obtain the allowable bending moments listed below for cases (a) through (d).

(a) $M = 6{,}750$ lbf-in (human bone)   (b) $M = 4{,}500$ lbf-in (pine)
(c) $M = 27{,}000$ lbf-in (steel)   (d) $M = 180$ lbf-in (concrete)

The allowable forces are obtained by dividing the moments by the length of the beam, $L = 15$ in.

(a) $F = 450$ lbf (human bone)   (b) $F = 300$ lbf (pine)
(c) $F = 1{,}800$ lbf (steel)   (d) $F = 12$ lbf (concrete)

The result indicates that the bone in your arm can resist a weight comparable to your own weight as described in Figure 2.5.7a.[*] Using the approach described above, you can also calculate the maximum wind force that a pine tree can resist.

The force that the beam can resist in axial tension can be obtained from $T = A\sigma$, where $A = bh = 1.8$ in.^{2}

(a) $T = 27{,}000$ lbf (human bone)   (b) $T = 18{,}000$ lbf (pine)
(c) $T = 108{,}000$ lbf (steel)   (d) $T = 720$ lbf (concrete)

Note that the ratio of the force that can be resisted in direct tension to that in flexure is $AL/Z = 60$ for the beam considered.

---

[*] The actual bone is not as simple as assumed: it is a pipe structure that is soft inside. As we have seen in this section, the center of the beam carries smaller stress and therefore need not be strong.

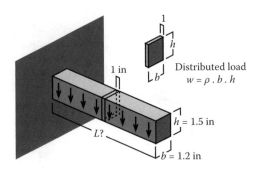

FIGURE 2.5.8 Limiting length of span under self-weight.

EXAMPLE 2.5.2

Calculate the limiting length of a beam having the same section and strength as in the previous example. In this case assume that the only load is the self-weight of the beam. Assume the specific weight to be:

(a) 0.04 lbf/in³ (human bone) (b) 0.015 lbf/in³ (pine)
(c) 0.3 lbf/in³ (steel) (d) 0.08 lbf/in³ (concrete)

Solution

The distributed load for each material is:

(a) $w = 0.072$ lbf/in (human bone) (b) $w = 0.027$ lbf/in (pine)
(c) $w = 0.54$ lbf/in (steel) (d) $w = 0.144$ lbf/in (concrete)

On the other hand, the bending moment caused by the distributed load, w, is $M = wL^2/2$ at the fixed end. Using the allowable bending moments obtained in the previous example, we have the following results.

(a) $L = \sqrt{\frac{2M}{w}} \sqrt{\frac{2 \times 6{,}750}{0.072}} = 433$ in (human bone)

(b) $L = \sqrt{\frac{2M}{w}} = \sqrt{\frac{2 \times 4{,}500}{0.027}} = 577$ in (pine)

(c) $L = \sqrt{\frac{2M}{w}} = \sqrt{\frac{2 \times 27{,}000}{0.54}} = 316$ in (steel)

(d) $L = \sqrt{\frac{2M}{w}} = \sqrt{\frac{2 \times 180}{0.144}} = 50$ in (concrete)

You may note the impressive strength of bio-materials. Reviewing the list above, you may consider concrete to be inferior to steel or timber, but concrete is in fact a good structural material if reinforced with steel or subjected to compression.

EXAMPLE 2.5.3

Repeat Example 2.5.1 assuming that both the width and the depth of the section are doubled. Will the limiting length of span also be doubled?

Moments and Deflections in Cantilever Beams

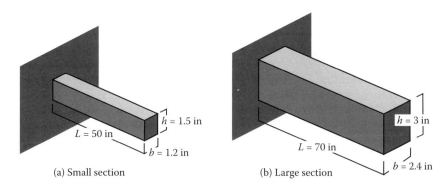

FIGURE 2.5.9 Beams with allowable length.

Solution

The section modulus $Z = bh^2/6$ will be $2^3 = 8$ times of that for the section shown in Figure 2.5.9a. Therefore, the allowable moment also will be 8 times as much. On the other hand, the distributed load will be $2^2 = 4$ times as much. Because L is proportional to $\sqrt{M/w}$, L shall be $\sqrt{8/4} = \sqrt{2}$ times as shown in Figure 2.5.9b. Note that the (length)/(depth) ratio decreases as the beam depth increases. This is the reason why the legs of elephants are so stocky compared with those of mice.

Design your own beam (Part 4)

We want to design a beam to carry a mini elephant, with weight equal to (any one digit number you choose) plus 10 lbf. Assume that each of the four legs carries the same amount of gravity force. The density of the beam is 0.5 lbf/in³. The beam width is 1 in. The tensile strength of the material is 500 psi. What is the required beam depth

(a) if the beam is prismatic (its section remains the same along its span) (Figure 2.5.10a), and
(b) if it has varying depth as specified in Figure 2.5.10b?

Check your results using GOYA-C.

Hint: The stress $\sigma = M/Z$, will be a maximum at the fixed end. The condition $M/Z = 500$ psi will lead to quadratic equations for the depth h.

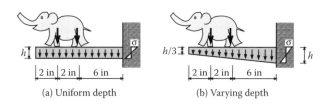

FIGURE 2.5.10 Beams to carry elephant.

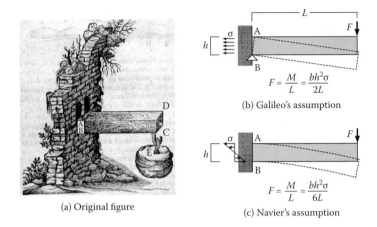

FIGURE 2.5.11 Galileo's cantilever beam.

Coffee break

In this section, you have learned that bending stress varies linearly over the depth of a beam in the linear range of response. You also know that the section subjected to bending only has an axis (*neutral axis*) where the stress is zero and that this axis is at mid-depth of a rectangular section. Recognizing this simple fact, however, has required many centuries of thinking. The first scientist who studied the strength of a cantilever beam was Galileo Galilei (1564–1642) (Figure 2.5.11a). He assumed that the bending stress was constant over the depth of the section as shown in Figure 2.5.11b. After him, the stress distribution in a bending beam was investigated by a lot of prominent scientists and mathematicians such as Bernoulli, Euler and Coulomb in the 17th and 18th centuries. Finally, a book published in 1826 by Navier (1785–1836) put an end to the controversy about the stress distribution (Figure 2.5.11c). [See 'History of Strength of Materials' by S.P. Timoshenko.]

Galileo Galilei, 1638. *Dialogues Concerning Two New Sciences*, Translated by H. Crew and A. de Salvio, 1954. Dover Publications, New York.

2.6 THE EFFECTS OF MOMENT ON STRAINS IN A BEAM

Consider a cantilever beam subjected to a couple M at its free end (Figure 2.6.1a). We shall investigate the portion of the beam between two planes (P_1 and P_2 in Figure 2.6.1a). Before the application of the moment, the two planes are perpendicular to the beam axis, which is straight. We assume the two planes to remain plane and perpendicular to the deflected beam axis after the beam is deformed in bending* (Figure 2.6.1b). The portion between the two planes is shown in detail in Figure 2.6.2.

---

* This assumption is valid only for pure bending but can be used to arrive at plausible results except for beams with span to depth ratios larger than two.

Moments and Deflections in Cantilever Beams

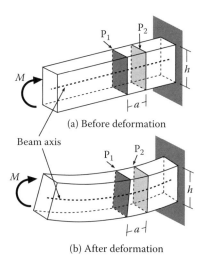

(a) Before deformation

(b) After deformation

FIGURE 2.6.1 Planes P_1 and P_2.

We shall redraw Figure 2.6.2 to make the *y*-axis vertical as shown in Figure 2.6.3a. Given the angle θ between P_1 and P_2, we infer that the top of the beam shortens by $\theta.h/2$ and the bottom lengthens by the same amount as shown in Figure 2.6.3b. As defined in Chapter 1, the changes in length shown in Figure 2.6.3b divided by the original length *a* lead to the strain distribution shown in Figure 2.6.3c, where the term

$$\phi = \frac{\theta}{a} \tag{2.6.1}$$

is introduced. The term ϕ is called the 'unit curvature' because it represents the amount that the beam axis curves over a unit length.* Noting that the strain is negative (compression) at the top and positive (tension) at the bottom, the strain distribution can be illustrated as shown in Figure 2.6.3d as a linear function of *y*:

$$\varepsilon = -\phi \cdot y \tag{2.6.2}$$

For a material that responds linearly, the relationship between stress and strain is written as

$$\sigma = E \cdot \varepsilon \tag{1.2.3}$$

where *E* is Young's modulus. Substituting $\varepsilon = -\phi \cdot y$ into Equation 1.2.3, we have

$$\sigma = -E \cdot \phi \cdot y \tag{2.6.3}$$

---

* In Figure 2.6.3a, *r* denotes the radius of the beam axis in its bent form. Because $r.\theta = a$ and $\phi = \theta/a$, *r* is a reciprocal of the unit curvature ($r = 1/\phi$), and is called the 'radius of curvature.'

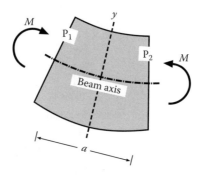

FIGURE 2.6.2 Detail of region between Planes P_1 and P_2.

As discussed in Section 2.5, bending moment is

$$M = -\int_{-h/2}^{h/2} \sigma \cdot y \cdot b \cdot dy \qquad (2.5.3)$$

Substituting Equation 2.6.3 into Equation 2.5.3, we can develop a relationship between bending moment and unit curvature for a rectangular section:

$$M = E \cdot \phi \cdot b \cdot \int_{-h/2}^{h/2} y^2 dy = E \frac{bh^3}{12} \phi$$

or

$$M = EI\phi \qquad (2.6.4)$$

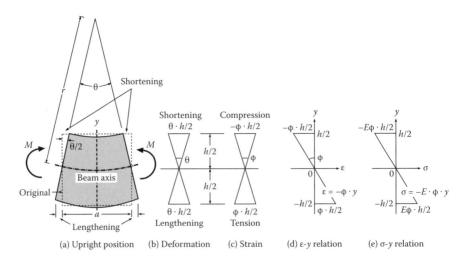

FIGURE 2.6.3 Deformation and strain between P_1 and P_2.

Moments and Deflections in Cantilever Beams

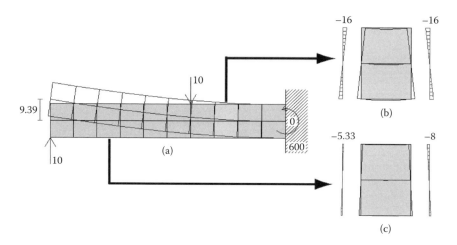

FIGURE 2.6.4 Strain distribution ($\times 10^{-3}$).

where

$$I = \frac{bh^3}{12} \tag{2.6.5}$$

The term I is called the 'moment of inertia'.[*] Although Equation 2.6.4 was derived for the special case of uniform bending moment, it is also applicable if the bending moment varies with x. In such cases, we obtain the same equation assuming the bending moment to be constant within an infinitesimal length dx.

The initial setting of GOYA-C is $b = 10$ mm and $h = 15$ mm, which leads to the following moment of inertia:

$$I = \frac{bh^3}{12} = \frac{10 \times 15^3}{12} = 2810 \text{ mm}^4$$

Apply forces as shown in Figure 2.6.4a to have a bending moment of 600 N-mm and click 'Strain' to obtain the strain distribution. Young's modulus is assumed to be 100 N/mm² in GOYA-C. The unit curvature caused by the bending moment of 600 N-mm is

$$\phi = \frac{M}{EI} = \frac{600}{100 \times 2810} = 2.13 \times 10^{-3}/\text{mm}$$

Note that the unit curvature has the unit of (1/length), because it is the inverse of the radius. Substituting the unit curvature determined above into $\varepsilon = -\phi \cdot y$, we

---

[*] The reason for the designation "moment of inertia" will be discussed in Chapter 4.

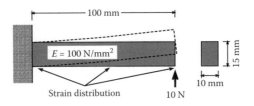

FIGURE 2.6.5 A force at the free end.

obtain the strain at the top of the beam ($y = 7.5$ mm) as

$$\varepsilon = -(2.13 \times 10^{-3}/\text{mm}) \times (7.5\text{mm}) = -16 \times 10^{-3} \quad \text{(compression)}$$

and that at the bottom ($y = -7.5$ mm) as

$$\varepsilon = -(2.13 \times 10^{-3}/\text{mm}) \times (-7.5\text{mm}) = +16 \times 10^{-3} \quad \text{(tension)}$$

as shown in Figure 2.6.4b. These strains imply that each rectangle in the grid having a dimension of 10 mm shortens $\varepsilon \times 10$ mm $= 0.16$ mm at the top and lengthens 0.16 mm at the bottom. Next, click the grids between the two forces, where the bending moment is not constant (Figure 2.6.4c). You will find that the strain gets smaller as the bending moment gets smaller. Thus, the strain varies not only in direction y but also in direction x if the moment varies along the span of the beam.

EXAMPLE 2.6.1

Assume that $L = 100$ mm, $b = 10$ mm, $h = 15$ mm, and $E = 100$ N/mm², as in GOYA-C. Apply a point load of 10 N at the free end (Figure 2.6.5). Evaluate the distribution at three sections: (1) the fixed end, (2) mid-span and (3) the free end.

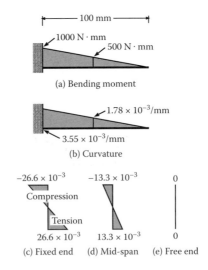

FIGURE 2.6.6 Strain distribution.

Moments and Deflections in Cantilever Beams

Solution

The moment of inertia is $I = 2810$ m⁴ as determined above. The bending moment is distributed as shown in Figure 2.6.6a. Dividing each ordinate of the moment distribution by $EI = 281 \times 10^3$ N/mm², we have the unit curvature distribution shown in Figure 2.6.6b. Using the expression $\varepsilon = -\phi \cdot y$, we get the strain distributions of shown in Figures 2.6.6c through e.

EXERCISE

Use any two digits i and j to change the load to $F = (i+2)$ N and the depth to $h = (j+5)$ mm, and solve the example above. Check your answer using GOYA-C.

2.7 DEFORMATION OF A BEAM—SPRING MODEL

In Section 2.6, we studied the strains in a beam using the continuously deformable model shown in Figure 2.7.1a. In this section, we shall simplify the model as shown in Figure 2.7.1b, where all the deformation is assumed to be concentrated (or "lumped") at the middle of the segment considered. In Figure 2.7.1b, a rotation

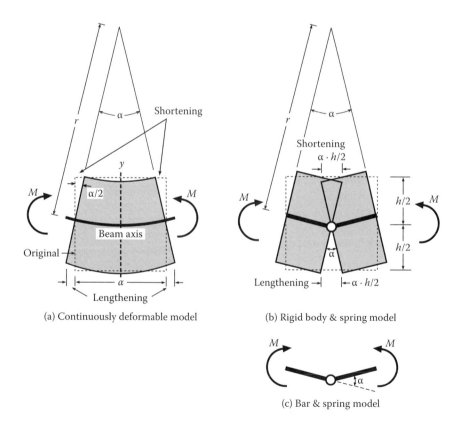

(a) Continuously deformable model

(b) Rigid body & spring model

(c) Bar & spring model

FIGURE 2.7.1 Spring model.

occurs at the middle of the segment considered but the two half segments remain rigid (un-deformed). Note that the total shortening and the lengthening are the same as those in the continuously deformable model. Using $\phi = \alpha/a$ and $M = EL\phi$, we obtain

$$\alpha = \frac{M}{EI} a \qquad (2.7.1)$$

We do this to develop a simple method to determine deflection as well as to understand the relationship between moment and deflected shape of a structural member. The representation in Figure 2.7.1b is essentially identical to the one shown in Figure 2.7.1c.

We shall use this approach to model a cantilever beam subjected to a concentrated force. Intuitively, we understand that the more springs we use, the better will be the result. If we assume ten springs as shown in Figure 2.7.2a, we may obtain a satisfactory result. However, try the two-spring model shown in Figure 2.7.2b to investigate if that will also give an acceptable result. The bending moments and the

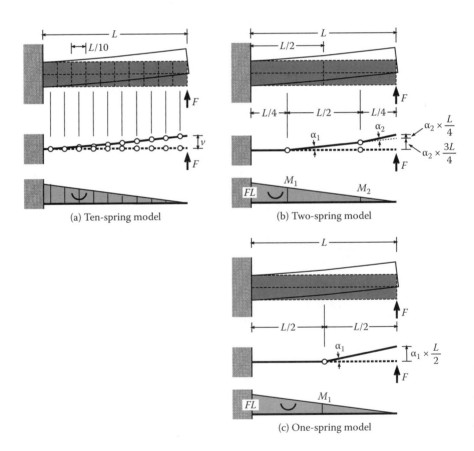

FIGURE 2.7.2 Spring model for a cantilever beam.

Moments and Deflections in Cantilever Beams

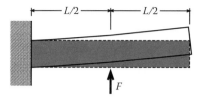

FIGURE 2.7.3 A load at the mid-span.

rotations at the springs are:

$$M_1 = F \times \frac{3}{4}L \qquad \alpha_1 = \frac{M_1}{EI} \times \frac{L}{2} = \frac{3FL^2}{8EI} \qquad (2.7.2)$$

$$M_2 = F \times \frac{1}{4}L \qquad \alpha_2 = \frac{M_2}{EI} \times \frac{L}{2} = \frac{FL^2}{8EI} \qquad (2.7.3)$$

The deflection of the free end is:

$$v = \alpha_1 \times \frac{3L}{4} + \alpha_2 \times \frac{L}{4} = \frac{9FL^3}{32EI} + \frac{FL^3}{32EI} = \frac{5FL^3}{16EI} \qquad (2.7.4)$$

This result is only 6% smaller than that given by the ten-spring model or by GOYA-C.* The exact result is $FL^3/3EI$.

What if we assume only one spring as shown in Figure 2.7.2c? The results are:

$$M_1 = \frac{FL}{2} \qquad \alpha_1 = \frac{M_1}{EI} \times L = \frac{FL^2}{2EI} \qquad (2.7.5)$$

$$v = \alpha_1 \times \frac{L}{2} = \frac{FL^3}{4EI} \qquad (2.7.6)$$

The computed deflection is 30% smaller than that given by the continuous model. This is not too bad.

EXERCISE

Use any two digits i and j to change the force to $F = (i + 2)$ N and the depth to $h = (j + 5)$ mm. Calculate the deflection at the free end using the two-spring model. Compare the result with that obtained by GOYA-C.

It is important to note that the deflection is inversely proportional to the moment of inertia I. If you change $b = 10$ mm and $h = 15$ mm to $b = 15$ mm and $h = 10$ mm (i.e. rotate the section by 90 degrees), $I = bh^3/12$ will be reduced by $1/1.5^2$ (approximately $1/2$) and the deflection will be approximately twice what it was. A shallow beam deflects more than a deep beam of comparable width supporting the same load.

---

* GOYA-C is based on the continuously deformable model which assumes infinite number of springs. This will be discussed in Section 2.8.

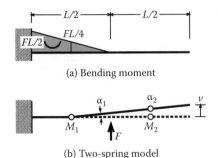

FIGURE 2.7.4 Spring model.

EXAMPLE 2.7.1

A force F is applied at mid-span of a cantilever beam. Calculate the deflection at the free end using the two-spring model.

Solution

According to Figure 2.7.4a, the bending moments and the rotations of the springs are:

$$M_1 = \frac{FL}{4} \qquad \alpha_1 = \frac{M_1}{EI} \times \frac{L}{2} = \frac{FL^2}{8EI} \qquad (2.7.7)$$

$$M_2 = 0 \qquad \alpha_2 = 0 \qquad (2.7.8)$$

The deflection of the free end is:

$$v = \alpha_1 \times \frac{3L}{4} + \alpha_2 \times \frac{L}{4} = \frac{3FL^3}{32EI} + 0 = \frac{3FL^3}{32EI} \qquad (2.7.9)$$

This result is 10% smaller than that given by the continuous model. Note that the deflection for this case is only 30% of that for the case of a load at the free end (Equation 2.7.4). Check this result using GOYA-C.

EXERCISE

What would the deflection be if we move the force to the left as shown in Figure 2.7.5?

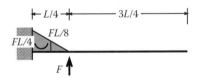

FIGURE 2.7.5 Move the force to the left.

Moments and Deflections in Cantilever Beams

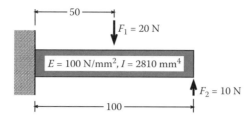

FIGURE 2.7.6 Loads at mid-span and at the free end.

Hint: Use of the two-spring model in Figure 2.7.4b results in zero deflection. You need to have the springs at different locations.

EXAMPLE 2.7.2

Two loads, F_1 and F_2, are to be applied at the free end and at mid-span of a cantilever beam as shown in Figure 2.7.6. Calculate the deflection at the free end using the two-spring model for $F_1 = 20$ N and $F_2 = 10$ N. (Use the default values in GOYA-C for Young's modulus and the moment of inertia.)

Solution

According to Figure 2.7.7a, the bending moments and the rotations of the springs are:

$$M_1 = M_2 = 250 \text{ N} \cdot \text{mm}$$

$$\alpha_1 = \alpha_2 = \frac{250}{100 \times 2810} \times 50 = 4.45 \times 10^{-2} \text{ rad}$$

The deflection of the free end is:

$$v = \alpha_1 \times 75 + \alpha_2 \times 25 = 4.45 \text{ mm}$$

Simulate this using GOYA-C and compare the results.

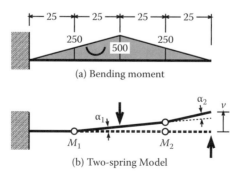

FIGURE 2.7.7 Solution.

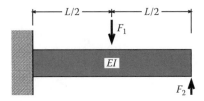

FIGURE 2.7.8 Loads at mid-span and at the free end.

EXERCISE

Take any two digits i and j. Assume $F_1 = (i + 2)$ N and $F_2 = (j + 5)$ N in Figure 2.7.6. Calculate the deflection at the free end. (Use the default values in GOYA-C for the size of the section and Young's modulus.)

EXAMPLE 2.7.3

Two loads, F_1 and F_2, are to be applied at the free end and at mid-span of a cantilever beam (Figure 2.7.8) so that the deflection at the free end is zero. Determine the ratio F_1/F_2 using the two-spring model.

Solution

Because the deflection at the free end must be zero, the beam shall deflect as shown in Figure 2.7.9b. This implies that the bending moment M_1 should be negative (concave downward) and M_2 should be positive (concave upward) as shown in Figure 2.7.9a. The moment equilibrium requires:

$$M_1 = F_1 \times \frac{L}{4} - F_2 \times \frac{3L}{4} = (F_1 - 3F_2) \times \frac{L}{4} \qquad (2.7.10)$$

$$M_2 = F_2 \times \frac{L}{4} \qquad (2.7.11)$$

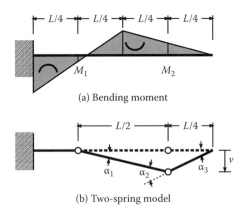

FIGURE 2.7.9 Solution.

Moments and Deflections in Cantilever Beams

The rotations of the springs are:

$$\alpha_1 = \frac{M_1}{EI} \times \frac{L}{2} = \frac{L^2}{8EI}(F_1 - 3F_2) \qquad (2.7.12)$$

$$\alpha_2 = \frac{M_2}{EI} \times \frac{L}{2} = \frac{L^2}{8EI} F_2 \qquad (2.7.13)$$

On the other hand, Figure 2.7.9b requires:

$$\alpha_1 = \frac{v}{L/2} = \frac{2v}{L}$$

$$\alpha_3 = \frac{v}{L/4} = \frac{4v}{L}$$

and

$$\alpha_2 = \alpha_1 + \alpha_3 = \frac{6v}{L} = 3\alpha_1$$

Substituting Equations 2.7.12 and 2.7.13 into $\alpha_2 = 3\alpha_1$ leads to:

$$F_2 = 3(F_1 - 3F_2)$$

or

$$\frac{F_1}{F_2} = \frac{10}{3} \qquad (2.7.14)$$

Figure 2.7.10 shows the result obtained from GOYA-C, where the green numeral (–3.9) indicates the maximum deflection in the span. If you click the segment at the

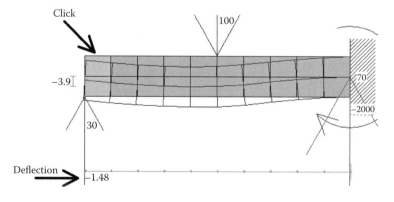

FIGURE 2.7.10 Simulation.

free end, the deflection of the free end (−1.48) will be indicated. If you increase F_1 from 30 to 31.25, the deflection of the free end will be exactly zero, as will be discussed in the next section.

EXAMPLE 2.7.4

Calculate the deflection of a cantilever beam with a length L subjected to a uniformly distributed load w pushing the beam up.

Solution

As we studied in Section 2.3 (see Figure 2.7.11a), the bending moment over the span is:

$$M = \frac{w}{2}x^2 \tag{2.3.3}$$

where x is a coordinate indicating the distance from the fixed end. The bending moments and the rotations of the springs are:

$$M_1 = \frac{9wL^2}{32} \qquad \alpha_1 = \frac{M_1}{EI} \times \frac{L}{2} = \frac{9wL^3}{64EI} \tag{2.7.15}$$

$$M_2 = \frac{wL^2}{32} \qquad \alpha_2 = \frac{M_2}{EI} \times \frac{L}{2} = \frac{wL^3}{64EI} \tag{2.7.16}$$

The deflection of the free end is:

$$v = \alpha_1 \times \frac{3L}{4} + \alpha_2 \times \frac{L}{4} = \frac{27wL^4}{256EI} + \frac{wL^4}{256EI} = \frac{7wL^4}{64EI} \tag{2.7.17}$$

The result is 12 % smaller than that given by the continuously deformable model. Note that the deflection for uniform load is proportional to the fourth power of the beam length.

The two-spring model is accurate enough for most practical uses.

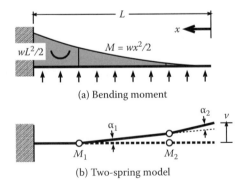

FIGURE 2.7.11 Distributed load.

2.8 DEFORMATION OF A BEAM—CONTINUOUSLY DEFORMABLE MODEL

If we differentiate a quadratic function such as

$$y = ax^2 + bx + c \qquad (2.8.1)$$

we get

$$dy/dx = 2ax + b \qquad (2.8.2)$$

As shown in Figure 2.8.1a, dy/dx represents the slope of the curve defined by the function. If we differentiate it again, we obtain

$$d^2y/dx^2 = 2a \qquad (2.8.3)$$

As shown in Figure 2.8.1b, the radius of the curve corresponding to Equation 2.8.3 is reduced as the constant a increases. If a is negative, the curve is bent in the opposite direction. If the curve is defined by trigonometric ($y = \sin x$) or exponential ($y = e^x$) functions, the term d^2y/dx^2 also indicates how the curve is bent.

The broken line in Figure 2.8.2a shows the deflected shape of the beam axis. The deflection at a distance x from the fixed end is denoted as v. Figure 2.8.2b shows the deflection v as a function of x, the distance from the fixed end. The first-order derivative of the deflection,

$$\theta = \frac{dv}{dx} \qquad (2.8.4)$$

represents the slope of the beam axis. As we did in Section 2.6, we shall again assume that the planes P_1 and P_2 in Figure 2.8.2a remain perpendicular to the beam axis, and

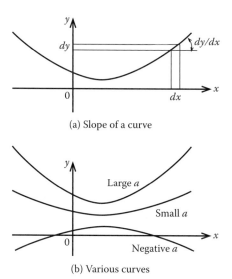

(a) Slope of a curve

(b) Various curves

FIGURE 2.8.1 Curves representing quadratic functions.

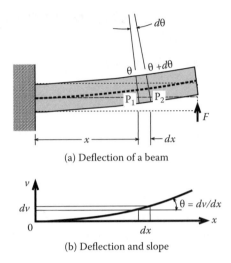

FIGURE 2.8.2 Cantilever beam.

define the unit curvature as

$$\phi = \frac{d\theta}{dx} \quad (2.8.5)$$

Note that Equation 2.8.5 is equivalent to Equation 2.6.1 ($\phi = \alpha/a$) except that the curvature is defined for infinitesimal length dx rather than a finite length a. Substituting Equation 2.8.4 into Equation 2.8.5 leads to:

$$\phi = \frac{d^2v}{dx^2} \quad (2.8.6)$$

In other words, the second-order differential of the deflection indicates how the beam bends* or how the slope changes over a very small distance ($d\theta/dx$).

---

* Equations 2.8.5 and 2.8.6 are approximate. The exact definition of unit curvature is

$$\phi = \frac{d\theta}{ds} \quad (2.8.7)$$

where ds is the length of the curve shown in Figure 2.8.3. The curve with a constant curvature is a circle with a radius $1/\phi$. Using calculus, it can be shown that:

$$\phi = \frac{d^2v}{dx^2} \bigg/ \left[1 + \left(\frac{dv}{dx}\right)^2\right]^{3/2} \quad (2.8.8)$$

In the case of beam deflection, dv/dx is very small. Therefore, $(dv/dx)^2$ is negligible in beam, and thus we can assume $\phi = d^2v/dx^2$.

Moments and Deflections in Cantilever Beams

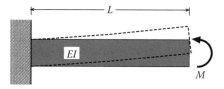

FIGURE 2.8.3 A couple at the free end.

Recalling Equation 2.6.4 ($M = EI\phi$), we have:

$$\frac{d^2v}{dx^2} = \frac{M}{EI} \tag{2.8.9}$$

This result should be compared with that of Equation 2.7.1, where the deformation of the beam is evaluated within a finite region a. Equation 2.8.9 deals with the bending deformation within an infinitesimal region dx. Solving this differential equation, we can determine the deflection formation of the beam. This is equivalent to assuming an infinite number of springs representing the deformation within an infinitesimal region dx.

EXAMPLE 2.8.1

Assume that a cantilever beam is subjected to a couple M at its free end ($x = L$). Determine the maximum slope and the maximum deflection of the beam.

Solution

First, we attempt to visualize how the beam would bend. The bending moment is distributed uniformly as shown in Figure 2.8.5a. The section does not change along the beam. We conclude that the curvature is uniform along the beam. Inspecting the sense of the applied moment, we conclude that the top fiber will be in compression and that the beam will bend into a concave upward shape. From this information, we infer that the beam will bend as sketched in broken lines in Figure 2.8.4 and that the maximum slope and deflection will be at the free end.

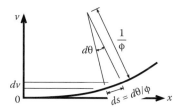

FIGURE 2.8.4 Definition of curvature.

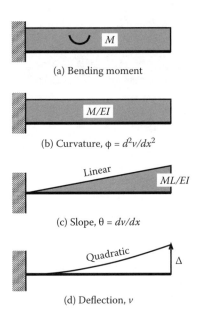

FIGURE 2.8.5 Deflection of a cantilever beam subjected to a couple.

We obtain the unit-curvature distribution using the expression $d^2v/dx^2 = M/EI$ (Figure 2.8.5b). Integrating this expression yields the slope

$$\theta = \frac{dv}{dx} = \frac{M}{EI} \times x + C_1$$

where C_1 is a constant of integration. Because the slope is zero at the fixed end ($dv/dx = 0$ at $x = 0$), we have $C_1 = 0$. The slope $\theta = dv/dx$ is distributed linearly along the span (Figure 2.8.5c). Integrating the expression for the slope and noting that the deflection is zero at the fixed end ($v = 0$ at $x = 0$), we get the following result (Figure 2.8.5d).

$$v = \frac{M}{EI} \times x^2$$

The slope and the deflection at the free end ($x = L$) are

$$\theta(L) = \frac{ML}{EI} \qquad (2.8.10)$$

and

$$\Delta = v(L) = \frac{ML^2}{EI} \qquad (2.8.11)$$

Moments and Deflections in Cantilever Beams

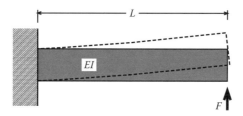

FIGURE 2.8.6 A force at the free end.

Exercise

Using two randomly selected numbers i and j change the couple in the previous example to $C = (i + 2) \times 100$ N-mm and the depth to $h = (j + 5)$ mm, and solve the example above again.

Example 2.8.2

Assume that a cantilever beam is subjected to a force F at the free end ($x = L$). Determine the slope and the deflection of the beam at the free end.

Solution

The bending moment varies linearly along the span as shown in Figure 2.8.7a:

$$M = F \times (L - x)$$

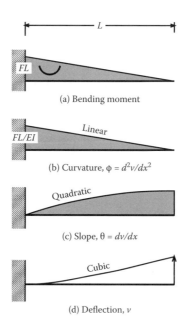

FIGURE 2.8.7 Deflection of a cantilever beam subjected to a force at the free end.

Use of $\phi = M/EI$ leads to the unit-curvature distribution shown in Figure 2.8.7b:

$$\frac{d^2v}{dx^2} = \frac{F}{EI}(L-x)$$

Integrating this equation and noting that the slope is zero at the fixed end ($dv/dx = 0$ at $x = 0$), we have

$$\theta = \frac{dv}{dx} = \frac{F}{EI}\left(Lx - \frac{x^2}{2}\right) \qquad (2.8.12)$$

The slope at the free end ($x = L$) is given by the following equation.

$$\theta(L) = \frac{FL^2}{2EI} \qquad (2.8.13)$$

Integrating Equation 2.8.12 and noting that the deflection is zero at the fixed end ($v = 0$ at $x = 0$), we have

$$v = \frac{F}{EI}\left(\frac{Lx^2}{2} - \frac{x^3}{6}\right)$$

The deflection at the free end ($x = L$) is given by the following equation.

$$\Delta = v(L) = \frac{FL^3}{3EI} \qquad (2.8.14)$$

If you substitute the default values used in GOYA-C ($F = 10$ N, $L = 100$ mm, $b = 10$ mm, $h = 15$ mm, and $E = 100$ N/mm^2), you should get $\Delta = 11.9$ mm. Check the results using GOYA-C.

EXERCISE

Use two numbers i and j you selected randomly to change the force to $F = (i + 2)$ N and the depth to $h = (j + 5)$ mm. Calculate the slope and deflection at mid-span and at the free end.

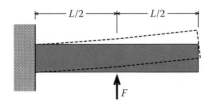

FIGURE 2.8.8 A load at mid-span.

Moments and Deflections in Cantilever Beams

EXAMPLE 2.8.3

A force F is applied at mid-span of a cantilever beam. Calculate the slope and deflection at the free end.

Solution

The bending moment varies linearly between $0 \leq x \leq L/2$ as shown in Figure 2.8.9a:

$$M = F \times \left(\frac{L}{2} - x\right) \quad (2.8.15)$$

Figure 2.8.9b shows the unit curvature, $\phi = M/EI$. Integrating this curvature and noting that the slope is zero at the fixed end ($dv/dx = 0$ at $x = 0$), we have

$$\theta = \frac{dv}{dx} = \frac{F}{EI} \times \left(\frac{Lx}{2} - \frac{x^2}{2}\right) \quad (2.8.16)$$

Integrating the expression for the slope and noting that the deflection is zero at the fixed end ($v = 0$ at $x = 0$), we have

$$v = \frac{F}{EI} \times \left(\frac{Lx^2}{4} - \frac{x^3}{6}\right) \quad (2.8.17)$$

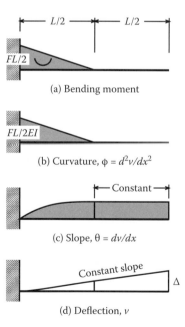

FIGURE 2.8.9 Deflection of a cantilever beam subjected to a force at mid-span.

The slope and the deflection at the mid-span of the beam ($x = L/2$) are

$$\theta\left(\frac{L}{2}\right) = \frac{FL^2}{8EI} \tag{2.8.18}$$

and

$$v\left(\frac{L}{2}\right) = \frac{FL^3}{24EI} \tag{2.8.19}$$

Note that we can obtain these equations substituting $L/2$ for L in Equations 2.8.13 and 2.8.14. Also note that the bending moment is zero in the right half of the beam ($L/2 \leq x \leq L$) as shown in Figure 2.8.9a. The unit curvature d^2v/dx^2 is also zero there, which leads to the conclusion that the slope dv/dx is constant in that region as shown in Figure 2.8.9c. Integrating Equation 2.8.18, we have

$$v = \frac{FL^2}{8EI} x + C \tag{2.8.20}$$

where C is a constant of integration. Substituting $x = L/2$ and Equation 2.8.19 into Equation 2.8.20, we can determine the constant C.

$$\frac{FL^3}{24EI} = \frac{FL^2}{8EI} \times \frac{L}{2} + C \quad \text{or} \quad C = -\frac{FL^3}{48EI} \tag{2.8.21}$$

Substituting $x = L$ and Equation 2.8.21 into Equation 2.8.20 leads to the deflection at the free end ($x = L$).

$$\Delta = v(L) = \frac{FL^2}{8EI} \times L - \frac{FL^3}{48EI} = \frac{5FL^3}{48EI} \tag{2.8.22}$$

EXAMPLE 2.8.4

Two loads, F_1 and F_2, are to be applied at the free end and at mid-span of a cantilever beam as shown in Figure 2.8.10. Determine the ratio F_1/F_2 required to make the deflection at the free end zero.

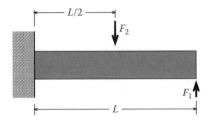

FIGURE 2.8.10 Loads at the free end and mid-span.

Moments and Deflections in Cantilever Beams

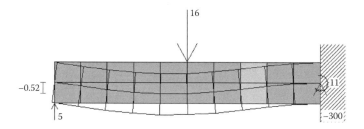

FIGURE 2.8.11 Two forces making the deflection at the free end zero (deformation magnified 8 times).

Solution

Though we can solve the problem by integrating the moment distribution created by F_1 and F_2, it is easier to use the results of the previous examples (Equations 2.8.14 and 2.8.22). Denoting the deflection at the free end created by F_1 and F_2 as Δ_1 and Δ_2, respectively, the deflection caused by the two forces is given by the following equation using Equations 2.8.14 and 2.8.22.

$$\Delta = \Delta_1 - \Delta_2 = \frac{F_1 L^3}{3EI} - \frac{5 F_2 L^3}{48 EI}$$

Setting $\Delta = 0$ leads to the following result

$$\frac{F_1}{F_2} = 3 \times \frac{5}{48} = \frac{5}{16}$$

Figure 2.8.11 shows the result obtained from GOYA-C.

EXERCISE

Take two numbers i and j you may choose. Assume $F_1 = (i + 2)$ N and $F_2 = (j + 5)$ N. Calculate the slope and deflection at mid-span and at the free end for the previous example. (Use the default values in GOYA-C for the size of the section and Young's modulus.)

EXERCISE

Take two numbers i and j you may choose. Assume the beam depth to be $h = (j + 10)$ mm. Calculate the set of forces that yields a maximum deflection of $\pm (i + 1)$ mm as shown in Figure 2.8.12. (Use the default values in GOYA-C for beam width and Young's modulus.)

Hint: Apply $F_1 = 10$ N in GOYA-C and increase F_2 from zero until the upward deflection equals the downward deflection.

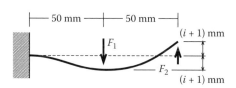

FIGURE 2.8.12 Deflected shape of cantilever for the loads shown.

EXAMPLE 2.8.5

Calculate the deflection of a cantilever beam with a length L subjected to a uniformly distributed load w pushing the beam up.

Solution

As we learned in Section 2.3, the shear force distribution can be determined by integrating the uniform load w over the span.

$$V = wx + C_1$$

If we define a coordinate indicating the distance from the fixed end, we get $C_1 = -wL$ because $V = 0$ at the free end ($x = L$).

Integrating the expression for shear, V, we obtain the bending moment distribution (Figure 2.8.13a).

$$M = \frac{w}{2}(x^2 - 2Lx) + C_2$$

We get $C_2 = w.L^2/2$ because $M = 0$ at the free end ($x = L$). The unit curvature ($\phi = M/EI$) is shown in Figure 2.8.13b or

$$\frac{d^2v}{dx^2} = \frac{w}{2EI}(x^2 - 2Lx + L^2) \qquad (2.8.23)$$

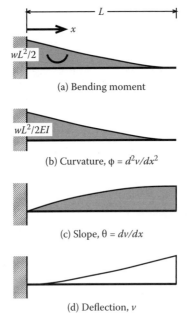

(a) Bending moment

(b) Curvature, $\phi = d^2v/dx^2$

(c) Slope, $\theta = dv/dx$

(d) Deflection, v

FIGURE 2.8.13 Deflection of a cantilever beam subjected to a distributed load.

Moments and Deflections in Cantilever Beams

Integrating the expression above and using the boundary conditions, we get Figure 2.8.13c or

$$\frac{dv}{dx} = \frac{w}{2EI}\left(\frac{x^3}{3} - Lx^2 + L^2x\right)$$

and Figure 2.8.13d or

$$v = \frac{w}{2EI}\left(\frac{x^4}{12} - \frac{Lx^3}{3} + \frac{L^2x^2}{2}\right)$$

The deflection at the free end is

$$\Delta = v(L) = \frac{wL^4}{8EI} \qquad (2.8.24)$$

EXERCISE

Take two number i and j you may choose. Assume the distributed force as $w = (i + 2)$ N/mm and the depth as $h = (j + 5)$ mm. Calculate the slope and deflection at mid-span and at the free end.

Mini-game using GOYA-C

The green Figures in the window show the maximum upward and downward deflections of the beam. Obtain the largest possible bending moment using three loads (Figure 2.8.14) while the maximum deflection does not exceed 1 mm. Keep the size of the beam and Young's modulus at the default values.

Design your own beam (Part 5)

We want to design a beam that can carry a nervous mini-elephant (Figure 2.8.15). The weight is (any one digit you choose plus 10 lbf. Assume that each leg carries the same amount of gravitational force. The density of the beam is 0.5 lbf/in³. The beam width is 1 in. and Young's modulus is 15,000 psi. Because the elephant is nervous, we are asked to design the beam so that its free end does not deflect more than 1 in. What is the required beam depth, h? Check your results using GOYA-C.

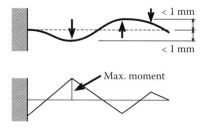

Major Structural League Ranking (Unit: N-mm)

> 200	> 500	> 1,000	> 2,000
Rookie	A	AA	3A
> 5,000	> 10,000	> 20,000	> 200,000
Major	All Star	MVP	Hall of Fame

FIGURE 2.8.14 Challange: use three loads to produce the largest bending moment without exceeding a deflection of 1 mm up and down.

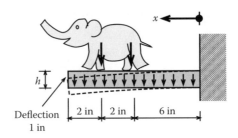

FIGURE 2.8.15 Uniform and concentrated loads on a cantilever.

Hint: You may use the equation obtained in Example 2.8.5. You will have to solve a cubic equation for the beam depth, h. Because the equation $(h^3 + ah^2 + bh + c = 0)$ is difficult to solve directly, you may resort to the following scheme: input $y = h^3 + ah^2 + bh + c$ into a spreadsheet and gradually increase h until you determine a deflection not exceeding 1 in.

What is unit curvature?

 Sir, I could not understand the idea of unit curvature.

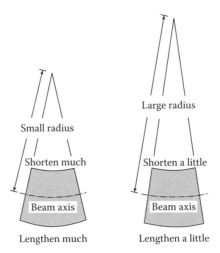

FIGURE 2.8.16 Large and small curvatures.

 OK. Imagine a circle which approximates the beam axis. Unit curvature is the inverse of the radius of the circle. A large curvature means a small radius, which means that the top of the beam shortens and the bottom lengthens considerably (Figure 2.8.16).

Moments and Deflections in Cantilever Beams

Another important point is $M = EI\phi$, which indicates that the unit curvature is proportional to the bending moment. You need more moment to bend the beam more.

The equation, $\phi = M/EI$, also has an important meaning. An oak tree is hard to bend because it has a high value of Young's modulus. A thick board is also hard to bend because it has large moment of inertia.

Yes, the wood I used in Chapter 1 bent very easily. But why are you so stiff when you are so slim?

Cough! Anyhow, unit curvature has another face: it is the second derivative of the deflection with respect to distance, $\phi = d^2v/dx^2$, which leads to the important equation.

$$\frac{d^2v}{dx^2} = \frac{M}{EI}$$

Integrating the bending moment, the beam deflection by calculation.

Unit curvature represents how a beam curves, and the beam deflects because of unit curvature. It sounds like a circular definition. By the way, I don't like calculus. It's boring. Could you tell me how to understand the idea intuitively?

Good question. In fact, calculus is rarely required in most structural design. Engineering intuition is much more important than calculus. The point is that a beam will be concave upward under positive bending moment and vice versa (Figure 2.8.17). You also need to train yourself so that you can imagine how the beam will deflect. Solve the problems

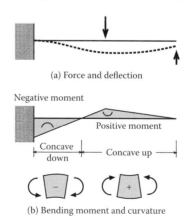

FIGURE 2.8.17 Deflected shape and bending moment.

in the next section and look at the graphics carefully order to improve your intuition for deflected shapes and their relation to bending-moment and shear-force distributions.

2.9 PROBLEMS

(Neglect self-weight in all the problems. Assume that all beams are prismatic.)

2.1 A cantilever beam is subjected to a couple M at its free end as shown in Figure 2.9.1. Select the correct pair of bending-moment distribution and free-end deflection from Table 2.9.1.

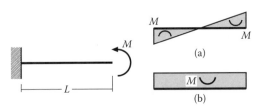

	Bending moment	Free-end deflection
1	(a)	0
2	(a)	$ML^2/(6EI)$
3	(a)	$ML^2/(3EI)$
4	(b)	0
5	(b)	$ML^2/(2EI)$

Table 2-9-1

FIGURE 2.9.1 Cantilever with a moment applied at its free end.

2.2 Two cantilever beams, one twice as long as the other, with identical sections are connected by a hinge at their free ends (Figure 2.9.2). A force F is applied at the hinge. Select the correct ratio of the vertical reactions R_A and R_C from Table 2.9.2.

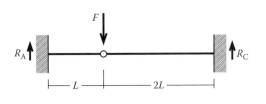

Table 2-9-2

	R_A/R_C
1	2
2	3
3	4
4	6
5	8

FIGURE 2.9.2 Two cantilevers meeting at a hinge

2.3 Two cantilever beams A and B, made of the same material, are loaded uniformly as shown in Figure 2.9.3. Select the correct ratio of their deflections at their free ends from Table 2.9.3.

Moments and Deflections in Cantilever Beams 157

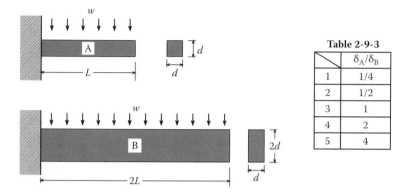

FIGURE 2.9.3 Uniformly loaded cantilevers with different spans and dimensions.

2.4 A load of 16 N is applied on the propped cantilever beam shown in Figure 2.9.4. Select the correct bending-moment diagram from Table 2.9.4.

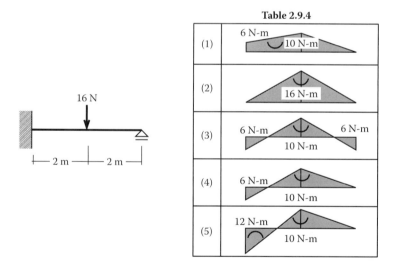

FIGURE 2.9.4 Propped cantilever with a concentrated load of 16 N at mid-span.

2.5 A couple of 10 N-m is applied on the pin-supported end of the propped cantilever beam shown in Figure 2.9.5. Select the correct bending-moment diagram for the beam from Table 2.9.5.

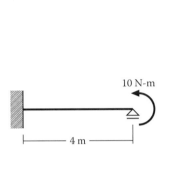

Table 2.9.5

(1)		10 N-m
(2)		10 N-m
(3)	5 N-m	10 N-m
(4)	10 N-m	10 N-m
(5)	15 N-m	10 N-m

FIGURE 2.9.5 Propped cantilever with a counterclockwise moment applied at its simple support.

2.6 A distributed load of 4 N/m is applied on the propped cantilever beam shown in Figure 2.9.6. Select the correct shear-force diagram for the beam from Table 2.9.6.

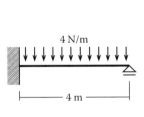

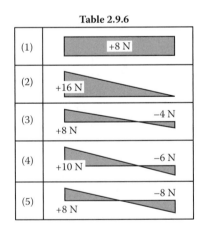

Table 2.9.6

(1)	+8 N
(2)	+16 N
(3)	+8 N / −4 N
(4)	+10 N / −6 N
(5)	+8 N / −8 N

FIGURE 2.9.6 Propped cantilever with uniform load.

2.7 Two columns made of the same linearly elastic material are connected with a rigid bar and a horizontal force is applied as shown in Figure 2.9.7. Select the correct ratio of the stresses at points a and b from Table 2.9.7.

Moments and Deflections in Cantilever Beams

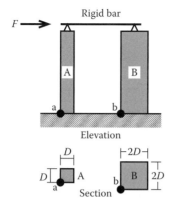

Table 2.9.7

	σ_a/σ_b
1	1/4
2	1/2
3	2
4	4
5	8

FIGURE 2.9.7 Two cantilever columns connected with a rigid for and subjected to a horizontal load at their free ends.

2.8 Assume that a cantilever beam with constant width but linearly varying depth is subjected to its self weight (Figure 2.9.8). Select the correct ratio of the bending moment at the fixed end M_A and that at mid-span M_B from Table 2.9.8.

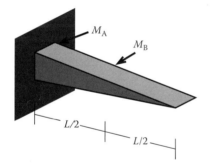

Table 2.9.8

	M_A/M_B
1	2
2	4
3	6
4	8
5	12

FIGURE 2.9.8 Cantilever with varying depth.

3 Moments and Deflections in Simply Supported Beams

3.1 THE EFFECT OF A CONCENTRATED LOAD ON SHEAR, MOMENT, AND DEFLECTION

Start GOYA-S to find a window showing a simply supported beam (Figure 3.1.1). This window shows a beam with its left and right ends supported by a pin and a roller, respectively. A perspective view of the beam is shown in Figure 3.1.2. Because the force F is located at midspan, the reactions from the supports are $F/2$.

In the lower right-hand corner of the screen, you will find a window showing the shear-force distribution (Figure 3.1.3). The shear force changes abruptly at the load point (midspan). You can see the reason for the change using Figure 3.1.4—if you cut the beam to the left of the load (Figure 3.1.4a), you will obtain Figure 3.1.4b, showing that the shear force is clockwise or positive; if you cut it to the right of the load (Figure 3.1.4c), you will obtain Figure 3.1.4d, showing that the shear force is counterclockwise or negative. Recall that the shear force of a cantilever beam also changed abruptly at the points with concentrated loads (see Section 2.2, Chapter 2).

In the lower left-hand corner of the screen, you will find a window showing the bending moment distribution (Figure 3.1.5). The moment is zero at both ends and reaches its maximum at the load point (midspan in this case). You can see how the moment distribution is obtained if you cut the beam as shown in Figure 3.1.6a to get the free-body diagram in Figure 3.1.6b, which indicates that a section at a distance x from the left support is subjected to a bending moment of

$$M = \frac{Fx}{2} \tag{3.1.1}$$

Note that $M = 0$ at $x = 0$ and that M increases linearly as x increases. The moment is said to be positive because the top of the beam is compressed. We can arrive at the same equation even if we consider the part of the beam to the right of the cut (Figure 3.1.6c). Equilibrium of moment around any point on the section gives the following equation:

(moment at the section, M) + (moment by the external force, F)
+ (moment by the reaction, $F/2$) = 0,

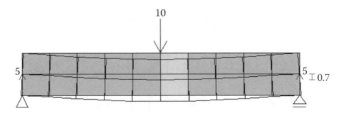

FIGURE 3.1.1 Window of a simply supported beam.

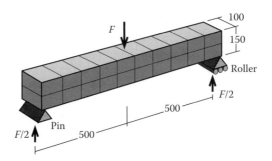

FIGURE 3.1.2 Perspective view of beam.

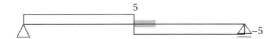

FIGURE 3.1.3 Shear-force window.

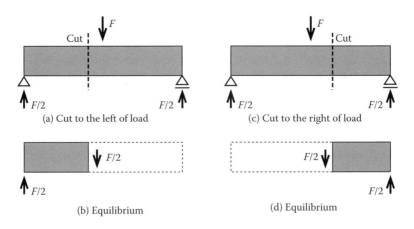

FIGURE 3.1.4 Cut the beam to see the shear force.

Moments and Deflections in Simply Supported Beams

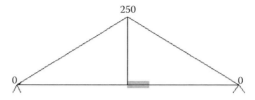

FIGURE 3.1.5 Bending moment.

or

$$M + F \cdot \left(\frac{L}{2} - x\right) - \frac{F}{2} \cdot (L - x) = 0$$

where clockwise moments are defined as being positive. This again results in $M = Fx/2$. If you cut the beam to the right of the load as shown in Figure 3.1.6d, you will obtain the free-body diagram shown in Figure 3.1.6e, which indicates that the

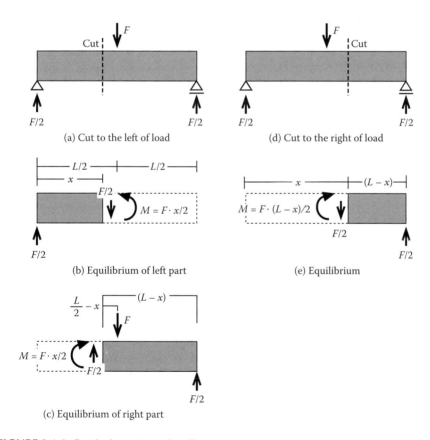

FIGURE 3.1.6 Cut the beam to see bending moment.

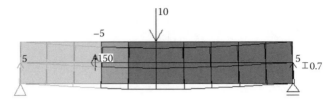

FIGURE 3.1.7 Cut the beam in the window.

bending moment is linearly proportional to the distance from the right support ($M = F.(L - x)/2$). The bending moment at the loaded point is, therefore,

$$M = \frac{FL}{2} \tag{3.1.2}$$

Click the free-body diagram button to obtain Figure 3.1.7, and confirm that the moment varies depending on the position.

EXAMPLE 3.1.1

Evaluate the force that would be required to break a chopstick of length 10 in. and section 0.2×0.2 in². The chopstick is supported simply and the force is applied at midspan. Assume that the tensile strength of the wood for the chopstick is 6000 psi.

Solution

As we learned in Section 2.4 the bending moment that causes a tensile stress of 6,000 psi is determined as follows.

$$M = \frac{bh^2}{6}\sigma = \frac{(0.2\,\text{in})}{6} \times (6,000\,\text{psi}) = 8\,\text{lbf-in}$$

Substituting the result for M into Equation 3.1.1, we obtain the force to break the chopstick:

$$P = \frac{4M}{L} = \frac{4 \times (8\,\text{lbf-in})}{10\,\text{in}} = 3.20\,\text{lbf}$$

This is a force that you can apply using your thumb. For comparison, the force required to break the chopstick by pulling is

$$P = bh\sigma = (0.2\,\text{in})^2 \times (6,000\,\text{psi}) = 240\quad\text{lbf}$$

or 75 times the force required if the chopstick is loaded as a simply supported beam.

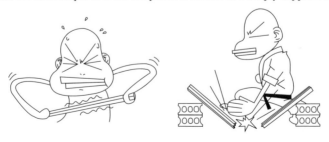

Moments and Deflections in Simply Supported Beams

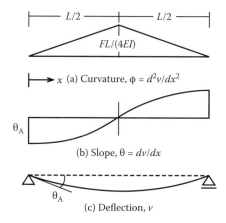

FIGURE 3.1.8 From curvature to slope and to deflected shape in a simply supported beam with a concentrated load at mid-span.

Next, we shall evaluate the deflection of a simply supported beam. Dividing the bending-moment distribution of Figure 3.1.5 by EI, we obtain the distribution of curvature shown in Figure 3.1.8a, or

$$\frac{d^2v}{dx^2} = \frac{Fx}{2EI} \quad \text{for} \quad 0 \le x \le L/2$$

Integrating this function we get Figure 3.1.8b or

$$\frac{dv}{dx} = \frac{Fx^2}{4EI} + \theta_A \tag{3.1.3}$$

where θ_A is a constant of integration representing the slope at the left end. Because the slope should be zero at midspan ($x = L/2$), we have

$$\theta_A = -\frac{FL^2}{16EI} \tag{3.1.4}$$

Integrating this equation with the boundary condition $v = 0$ at $x = 0$ leads to Figure 3.1.8c or

$$v = \frac{Fx^3}{12EI} + \theta_A x = \frac{Fx}{48EI}(4x^2 - 3L^2) \quad \text{for} \quad 0 \le x \le L/2 \tag{3.1.5}$$

The deflection in the right half ($L/2 \le x \le L$) can be obtained by replacing x with $(L - x)$ as shown in the following equation:

$$v = \frac{F(L-x)}{48EI}[4(L-x)^2 - 3L^2] \quad \text{for} \quad L/2 \le x \le L \tag{3.1.6}$$

The deflection at midspan ($x = L/2$) is

$$v = -\frac{FL^3}{48EI} \tag{3.1.7}$$

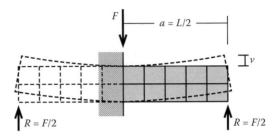

FIGURE 3.1.9 Equivalent cantilever beams.

where the negative sign represents that the beam deflects downward. Note that the deflection is proportional to the third power of the beam length L. Go back to GOYA-S. Push the "Detail of beam" button and double the beam length; the deflection will be $2^3 = 8$ times.

If you substitute $a = L/2$ and $R = F/2$ into Equation 3.1.7, you will have

$$v = -\frac{Ra^3}{3EI} \tag{3.1.7}$$

which is identical with the deflection of the cantilever beam (Section 2.8). The reason can be inferred from Figure 3.1.9—each half of the simply supported beam bends like a cantilever beam.

EXAMPLE 3.1.2

Evaluate the deflection at rupture for the chopstick discussed in Example 3.1.1. Assume that the Young's modulus is 1000 ksi.

Solution

As discussed in Section 2.5, Chapter 2, the moment of inertia of the section is

$$I = \frac{bh^3}{12} = \frac{0.2^4}{12} = 1.3 \times 10^{-4} \,(\text{in.}^4)$$

Substituting this into Equation 3.1.7 and noting that the force is 3.2 lbf, we obtain the deflection

$$v = -\frac{FL^3}{48EI} = -\frac{3.2 \times 10^3}{48 \times 1,000 \times 10^3 \times 1.3 \times 10^{-4}} \approx -0.5 \,(\text{in})$$

The strain when the chopstick breaks under pure tension is

$$\varepsilon = \frac{\sigma}{E} = \frac{6,000}{1,000 \times 10^3} = 0.006$$

The elongation at the break under pure tension is

$$e = \varepsilon \times L = 0.006 \times 10 = 0.06 \,(\text{in})$$

which is much smaller than the deflection of the simply supported beam.

Moments and Deflections in Simply Supported Beams

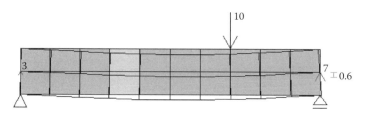

FIGURE 3.1.10 A simply supported beam with a load.

Exercise

Take any two numbers i and j you choose. Consider a timber beam of $L = 1$ m, $b = (5+i)$ mm, $h = (5+j)$ mm, tensile strength = 60 N/mm², and Young's modulus = 10,000 N/mm². Calculate the force and the deformation at rupture (a) if it is loaded transversely and (b) if it is loaded in tension axially.

Go back to GOYA-S and move the load to the right as shown in Figure 3.1.10. You will find that the reaction (green arrow and digit) at the right support increases, whereas the one at the left decreases as you move the load.

You can see the reason for the change of the reactions with the help of Figure 3.1.11a—if we consider the equilibrium of moments around the right support, we get the following equation with the clockwise moment defined as positive:

$$\Sigma M = R_A L - Fb = 0$$

which leads to

$$R_A = \frac{b}{L} F \qquad (3.1.8)$$

Similarly, the equilibrium of moments around the left support yields the following equation:

$$R_B = \frac{a}{L} F \qquad (3.1.9)$$

As you move the load to the right (or increase a), the reaction at the right support R_B increases. This is a demonstration of the principle of the lever.

In GOYA-S, look at the windows showing the bending moment and the shear force (Figure 3.1.12). The moment and shear diagrams also change as you move the force. If you cut the beam to the left of the applied load as shown in Figure 3.1.11b, you will find that

$$V = R_A \qquad (3.1.10)$$

and

$$M = R_A x \qquad (3.1.11)$$

If you cut the beam to the right of the applied load as shown in Figure 3.1.11c, you will find that

$$V = R_B \qquad (3.1.12)$$

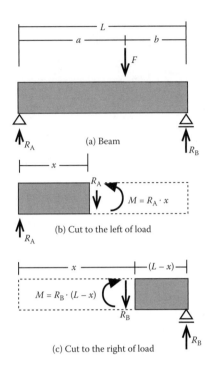

FIGURE 3.1.11 Free-body diagrams for a simply supported beam with a concentrated load.

and

$$M = R_B(L - x) \tag{3.1.13}$$

The maximum moment takes place at the loaded point, and its magnitude is

$$M_{max} = R_A a = \frac{ab}{L} F \tag{3.1.14}$$

M_{max} reaches its highest value if the load is placed at midspan ($a = b = L/2$).

Look at Figure 3.1.10 again. You may notice that the maximum deflection takes place *not* at the loaded point but near the middle of the beam. The deformation of

FIGURE 3.1.12 Bending-moment and shear-force diagrams in a simply supported beam with a concentrated load.

Moments and Deflections in Simply Supported Beams

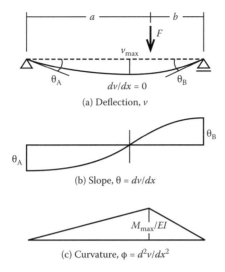

FIGURE 3.1.13 Deflected shape, slope distribution, and curvature distribution in a simply supported beam with a load applied at a point away from mid-span.

the beam is illustrated in Figure 3.1.13a, where you should note that the maximum deflection (v_{max}) occurs at the point of zero slope ($dv/dx = 0$).

For interested readers: We can detect the point of zero slope as follows. In the left part of the beam ($0 \leq x \leq a$), we have the following equation for curvature (the left part of Figure 3.1.13c):

$$\frac{d^2v}{dx^2} = \frac{M}{EI} = \frac{Fbx}{EIL} \tag{3.1.15}$$

Integrating this equation leads to the left part of Figure 3.1.13b or

$$\frac{dv}{dx} = \theta_A + \frac{Fbx^2}{2EIL} \tag{3.1.16}$$

where θ_A is the slope at the left end. Noting that $v = 0$ at $x = 0$, we obtain the left part of Figure 3.1.13a or

$$v = \theta_A x + \frac{Fbx^3}{6EIL} \tag{3.1.17}$$

For the right segment of the beam ($a \leq x \leq L$), we obtain the following equation:

$$\frac{d^2v}{dx^2} = \frac{M}{EI} = \frac{Fa(L-x)}{EIL} \tag{3.1.18}$$

Integrating this equation leads to the right part of Figure 3.1.13b or

$$\frac{dv}{dx} = \theta_B - \frac{Fa(L-x)^2}{2EIL} \tag{3.1.19}$$

where θ_B is the slope at the right end. Integrating this and noting $v = 0$ at $x = L$, we obtain the right part of Figure 3.1.13a or

$$v = -\theta_B(L-x) + \frac{Fa(L-x)^3}{6EIL} \qquad (3.1.20)$$

The deflection (v) should be continuous at the load point ($x = a$); i.e., the deflection computed with Equation 3.1.17 should be equal to that computed with Equation 3.1.20. The slope (dv/dx) should also be continuous at the load point. These boundary conditions lead to a set of simultaneous equations for θ_A and θ_B:

$$\theta_A = -\frac{Fb(L^2 - b^2)}{6EIL} \qquad (3.1.21)$$

$$\theta_B = \frac{Fa(L^2 - a^2)}{6EIL} \qquad (3.1.22)$$

If we substitute Equation 3.1.21 into Equation 3.1.16, we get

$$\frac{dv}{dx} = -\frac{Fb}{2EIL}(L^2 - b^2 - 3x^2) \qquad (3.1.23)$$

The point of zero slope ($dv/dx = 0$) is

$$x = \sqrt{\frac{L^2 - b^2}{3}} \qquad (3.1.24)$$

This is the point where the maximum deflection occurs (Figure 3.1.13a).* Substitute $b = L/10$ into Equation 3.1.24, for example, and you will find $x \approx 0.57L$, indicating that the maximum deflection occurs near midspan. As you increase the load, however, the beam will break at the load point because the maximum bending moment occurs at that point.

EXERCISE

Assume a simply supported chopstick with $L = 10$ in., $b = h = 0.2$ in., tensile strength = 6000 psi, and Young's modulus = 1000 ksi (Figure 3.1.14). Take any number i and apply a force at a distance of i in. from the right end of the chopstick. Determine at what force the chopstick will break.

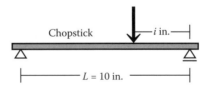

FIGURE 3.1.14 Simply supported chopstick.

---

* This equation is valid only for the case of $a \geq b$ because we assumed that the point of zero slope is located at the left of the load. You need to use Equation 3.1.19 to analyze the case of $a < b$.

3.2 THE EFFECT OF SEVERAL CONCENTRATED LOADS ON SHEAR, MOMENT, AND DEFLECTION

In GOYA-S, add a load and move it to suit Figure 3.2.1. We can evaluate the reactions of this beam superimposing Figures 3.1.1 and 3.1.10. We can also evaluate them directly as follows: In accordance with the forces and reactions shown in Figure 3.2.2, we consider the equilibrium of moments around the left support to obtain (clockwise moments assumed to be positive)

$$\Sigma M = F_1 L_1 + F_2 L_2 - R_B L = 0 \tag{3.2.1}$$

Equation 3.2.1 can be rearranged to determine the reaction at support B.

$$R_B = \frac{F_1 L_1 + F_2 L_2}{L} \tag{3.2.2}$$

We can similarly evaluate the reaction at the right support, R_A. Knowing the reactions, we can calculate the shear force, V. Noting that $dM/dx = V$, we obtain the bending moment, M.

EXERCISE

Take any number i you choose. Determine the forces that will produce the bending moments shown in Figures 3.2.3a,b. Check your results using GOYA-S and sketch the deformed shape of the beam.

Practice: Construct an interesting bending-moment diagram as you did in Section 2.2, Chapter 2.

Start GOYA-C and apply forces as shown in Figure 3.2.4 (the deformation is magnified by a factor of four). The force at the free end is the same as the reaction at the right in Figure 3.2.1. You will note that the equilibrium conditions for the cantilever beam and the simply supported beam are the same—including the bending-moment and shear-force diagrams. The only difference is provided by the boundary conditions. Let Δ denote the deflection at the free end. Rotate Figure 3.2.4 by Δ/L counter-clockwise. You will obtain the deformed shape shown in Figure 3.2.1.

Figure 3.2.5a shows a simply supported beam subjected to a couple at the left end. You should note that the bending-moment diagram (Figure 3.2.5b) is the same as that for the cantilever beam depicted in Figure 3.2.5c.

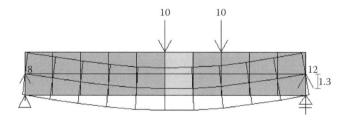

FIGURE 3.2.1 A simply supported beam with two loads.

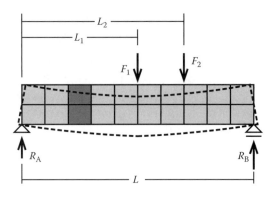

FIGURE 3.2.2 Reactions.

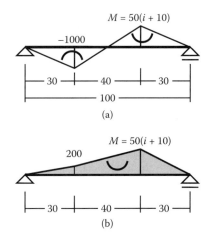

FIGURE 3.2.3 Bending moment.

Using GOYA-S, we can simulate Figure 3.2.5 as shown in Figure 3.2.6. In this case, the load of 60 N is applied at a distance of 5 mm from the left support, and the equivalent couple is $M = 60 \times 5 = 300$ N-mm; the reaction at the right support is $M/L = 300/100 = 3$ N.

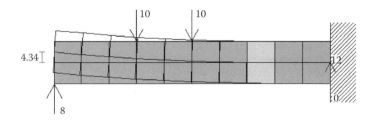

FIGURE 3.2.4 Bending-moment diagram of cantilever beam.

Moments and Deflections in Simply Supported Beams

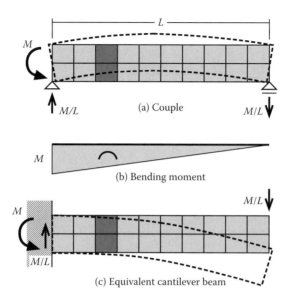

FIGURE 3.2.5 Simply supported beam with a couple at the left end.

Next, we calculate the shear force and the bending moment in the simply supported beam with a uniformly distributed load (Figure 3.2.7a). Because of symmetry, the reactions are

$$R_A = R_B = \frac{wL}{2} \tag{3.2.3}$$

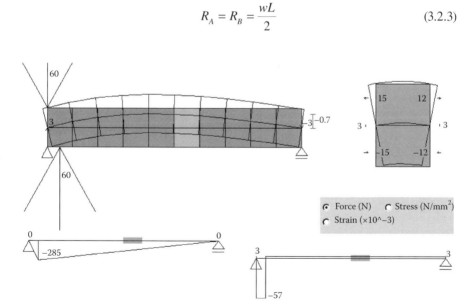

FIGURE 3.2.6 Bending-moment diagram similar to that of a cantilever beam.

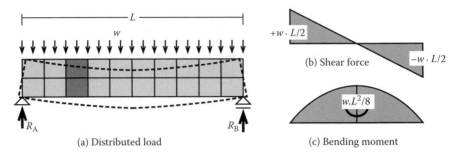

FIGURE 3.2.7 Simply supported beam under uniformly distributed load.

As we learned in Chapter 2, we have $dV/dx = -w$ (negative because the load w is downward). Noting that the shear force at the left support ($x = 0$) is $V = R_A$ (positive because it is clockwise), we have

$$V = R_A - wx = \frac{wL}{2}(L - 2x) \qquad (3.2.4)$$

The result is shown in Figure 3.2.7b. From Chapter 2 we know that $dM/dx = V$. Noting that the bending moment at the left support ($x = 0$) is zero, we write

$$M = R_A x + \frac{1}{2}wx^2 = \frac{wx}{2}(L - x) \qquad (3.2.5)$$

The result is shown in Figure 3.2.7c.

EXAMPLE 3.2.1

Calculate the deflection of the simply supported beam shown in Figure 3.2.7a.

Solution

Substituting Equation 3.2.5 into the equation $d^2v/dx^2 = M/EI$ (Chapter 2), we obtain an expression for curvature:

$$\frac{d^2v}{dx^2} = \frac{M}{EI} = -\frac{w}{2EI}(x^2 - Lx) \qquad (3.2.6)$$

Integrating Equation 3.2.6, we obtain an expression for the slope of the beam at any position x:

$$\frac{dv}{dx} = -\frac{w}{12EI}(2x^3 - 3Lx^2) + C_1 \qquad (3.2.7)$$

where C_1 is a constant of integration. Integrating Equation 3.2.7, we arrive at the expression for the deflection

$$v = -\frac{w}{24EI}(x^4 - 2Lx^3) + C_1 x + C_2 \qquad (3.2.8)$$

Moments and Deflections in Simply Supported Beams

where C_2 is another constant of integration. Knowing the two boundary conditions (the deflections at the left and the right supports are zero) helps solve the constants C_1 and C_2. Thus, we obtain

$$v = -\frac{w}{24EI}(x^4 - 2Lx^3 + L^3 x) \tag{3.2.9}$$

The deflection at midspan ($x = L/2$) is

$$v = -\frac{5wL^4}{384EI} \tag{3.2.10}$$

We can also derive this equation by considering the cantilever beam as having a length equal to half the length of the simply supported beam ($a = L/2$), as illustrated in Figure 3.2.8. The results in Chapter 2 indicate that the deflection of the cantilever beam of length a subjected to uniformly distributed load w is (see Figure 3.2.8b)

$$v_a = -\frac{wa^4}{8EI} \tag{3.2.11}$$

The deflection caused by an upward (reaction) force of wa is (see Figure 3.2.8c)

$$v_b = \frac{wa^4}{3EI} \tag{3.2.12}$$

The deflection caused by both loads is (see Figure 3.2.8a)

$$v = v_a + v_b = \frac{5wa^4}{24EI} \tag{3.2.13}$$

Substituting $a = L/2$ into Equation 3.2.13 leads to the same deflection as Equation 3.2.10 in magnitude.

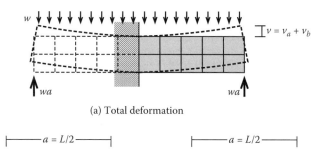

(a) Total deformation

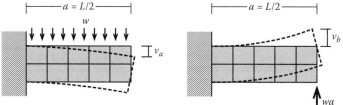

(b) Deflection caused by distributed load   (c) Deflection caused by reaction

FIGURE 3.2.8 Deflection of cantilever beam.

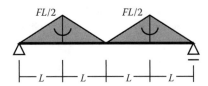

FIGURE 3.2.9 Bending-moment distribution.

EXAMPLE 3.2.2

From among the sets of forces shown in Figure 3.2.10(a–d), select the correct set that produces the bending-moment distribution shown in Figure 3.2.9. Select the correct deflected shape from among those shown in Figure 3.2.11(a–d) for the bending-moment distribution shown in Figure 3.2.9.

Solution

Applying $dM/dx = V$ for the moment distribution in Figure 3.2.9, we obtain the shear force shown in Figure 3.2.12. Because the changes in the shear force are caused by external forces, the correct force set is in Figure 3.2.10a. To find the correct deflection pattern, we need to recall that the bending-moment diagram (because of our sign convention) indicates whether the top or the bottom of the beam is compressed. Because Figure 3.2.9 indicates that the top is compressed everywhere in the beam, the correct answer must be that shown in Figure 3.2.11a.

We can simulate the result using GOYA-S. First, apply two downward forces as shown in Figure 3.2.13. Then, apply another force at the middle and increase its magnitude to –10 N, as shown in Figure 3.2.14, where the deformation is amplified 16 times.

If you increase the force to –13.8 N, you will find that the deflection at the middle will become zero, as shown in Figure 3.2.15, where the deformation is amplified 32 times. This is the deflected shape of Figure 3.2.11c. The beam does not bend abruptly, as shown schematically in Figure 3.2.11d.

Note that the deflected shape shown in Figure 3.2.15 is symmetric about midspan. In Example 2.5.5, for a cantilever beam with two loads applied at the free end and midspan (P_1 and P_2), we saw that the deflection at the free end is zero if $P_1/P_2 = -5/16$. In Figure 3.2.15, we see that the ratio of the reaction (3.1 N) to the downward force (–10 N) is $-3.1/10 \approx -5/16$.

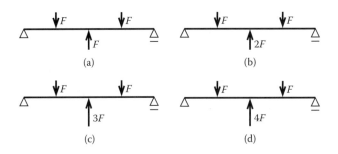

FIGURE 3.2.10 Loads.

Moments and Deflections in Simply Supported Beams

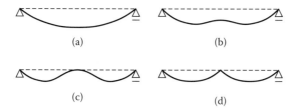

FIGURE 3.2.11 Deflection.

FIGURE 3.2.12 Shear force.

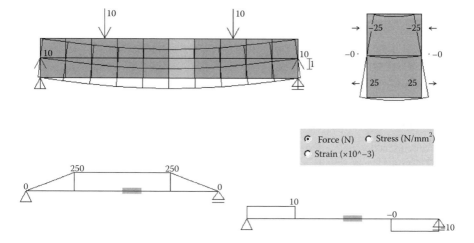

FIGURE 3.2.13 Two downward forces (deformation amplified four times).

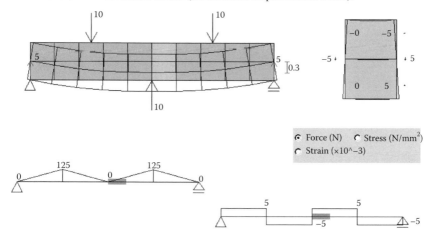

FIGURE 3.2.14 Upward force of −10 N (deformation amplified 16 times).

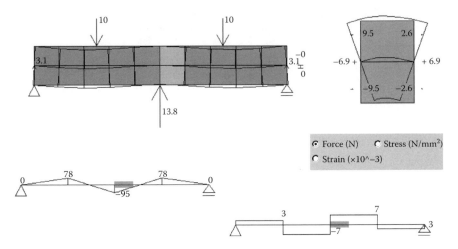

FIGURE 3.2.15 Upward force of −13.8 N (deformation amplified 32 times).

EXAMPLE 3.2.3

Calculate the shear force and the bending moment at A in the beam shown in Figure 3.2.16.

Solution

We represent the total uniform load of $6wL$ by a concentrated load located at the centroid of the uniform load (Figure 3.2.17a). The equilibrium of moment around point C gives

$$R_B = 6wL \times L/4L = 3wL/2$$

and the equilibrium of forces in the vertical direction gives

$$R_C = 6wL - R_B = 9wL/2$$

Next, we shall cut the beam at A as shown in Figure 3.2.17b and replace the distributed load by a concentrated load* of $2wL$. From the equilibrium of forces, we get

$$V_A = -wL/2$$

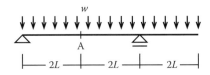

FIGURE 3.2.16 Simply supported beam with overhang.

---

* When you want to calculate the reactions, you may replace all the distributed load with an equivalent single load. When you want to calculate the shear force or the bending moment at a point, however, you need to go through the following procedure:
(a) Evaluate the reactions (Figure 3.2.17a).
(b) Cut the beam and *then* replace the distributed load with an equivalent single load (Figure 3.2.17b).
(c) Consider equilibrium of forces and moments.

Moments and Deflections in Simply Supported Beams

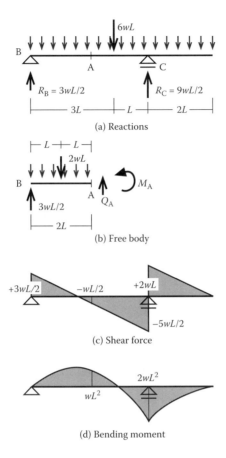

(a) Reactions

(b) Free body

(c) Shear force

(d) Bending moment

FIGURE 3.2.17 Determining shear-force and bending-moment diagrams in a beam with an overhanging span.

and from the equilibrium of moments, we get

$$M_A = wL^2.$$

We can similarly calculate the shear force and the bending moment at other sections. The results are shown in Figure 3.2.17c, d.

EXAMPLE 3.2.4

Calculate the bending moment at A in the beam shown in Figure 3.2.18.

Solution

Because the total beam length is $4L$, the total vertical load is $4wL$. Recognizing symmetry, we determine each reaction to be $2wL$, as shown in Figure 3.2.19a. Next, we cut the beam at A as shown in Figure 3.2.19b and replace the distributed load by a concentrated load of $2wL$. The moment equilibrium gives $M_A = 0$. We can similarly calculate the shear force and the bending moment at other places. The results are shown in Figure 3.2.19c, d.

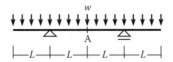

FIGURE 3.2.18 Uniformly loaded beam with two overhanging span.

Minigame Using GOYA-S

The green figures in the GOYA window show the maximum upward and downward deflections of the beam. Try producing the largest possible bending moment using four loads while controlling the maximum deflections so that they do not exceed 1 mm. Use default values for the size of the beam and Young's modulus.

Design Your Own Beam (Part 6)

We want to design a beam that can carry a mini-elephant whose weight is any number you choose plus 10 lbf. Assume that each leg carries the same amount of gravitational force. The density of the beam is 0.5 lbf/in.³, the beam width is 1 in., and

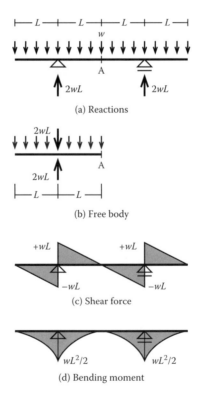

FIGURE 3.2.19 Determining shear-force and bending-moment diagrams in a beam with an overhanging span.

Moments and Deflections in Simply Supported Beams

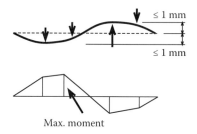

Max. moment

Major Structural League Ranking (N · mm)

> 100	> 200	> 500	> 1,000
Rookie	A	AA	3A

> 2,000	> 5,000	> 10,000	> 50,000
Major	All Star	MVP	Hall of Fame

FIGURE 3.2.20 Challenge: use four loads to produce the largest bending moment without exceeding a deflection of 1 mm up and down.

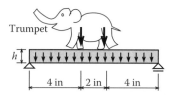

FIGURE 3.2.21 Uniformly loaded beam with elephant.

Young's modulus is 10 ksi. The elephant is scared of flexible beams. The maximum deflection of the beam should not exceed 0.1 in. The beam material cannot resist tensile stress exceeding 300 psi. What is the required beam depth, h? Check your results using GOYA-S. (Hint: The beam is symmetric. You may replace it by a cantilever beam as shown in Figure 3.2.8.)

3.3 SIMILARITIES BETWEEN BEAM AND TRUSS RESPONSE

In this section we will investigate the similarities in the ways beams and trusses resist transverse loads. We do this to develop an improved perspective of the relationships between internal and external forces, a perspective that will help improve our understanding of structural response.

In Figure 3.3.1a, b, we compare the internal and external forces in a beam and a truss, both loaded with vertical forces of magnitude F located symmetrically about midspan. We choose each of the 10 truss panels to have a length of 10 mm and a depth of 10 mm. Because the panels are square, the *web* members, or diagonals, make an angle of 45° with the horizontal. We choose a beam having the same span and a depth of 15 mm. Because of the symmetry, the magnitude of each reaction of the beam and the truss is determined to be F.

Determining the shear-force diagram for the beam is straightforward (Figure 3.3.1c). The shear force is constant between the left reaction and the force applied on the left. It is zero between the applied forces. Between the force applied on the right and the right reaction, the shear is again constant. The beam shear is equal in magnitude to the external force F and is shown to be positive on the left and negative on the right—to be consistent with our sign convention.

182 Understanding Structures: An Introduction to Structural Analysis

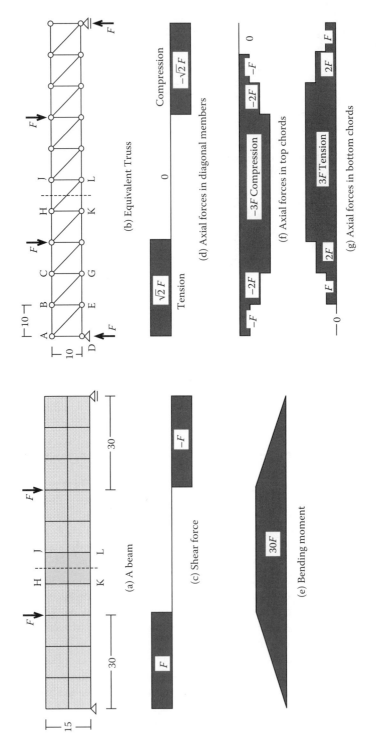

FIGURE 3.3.1 Internal forces in a beam (a) and an equivalent truss (b).

Moments and Deflections in Simply Supported Beams

The shear force in the truss is resisted by the web members (Figure 3.3.1d). There is a direct relation between the shear in the truss and the force in the web members. Because of the inclination of the web members (45°), the axial-force in each web member is $\sqrt{2}F$. The sign changes from the left to the right end of the truss because the web members work in tension on the left and in compression on the right.

Comparing the diagrams for shear distribution in the beam (Figure 3.3.1c) and force distribution in the web members of the truss (Figure 3.3.1d) we understand that there is a similarity as well as a proportionality between the internal shear distribution in a beam and the distributions of the forces in the web members of a truss.

Next, we examine the moment distribution in the beam (Figure 3.3.1e). We have studied the relation between shear and change in moment. So, it is not surprising to see there is a steady increase in moment in the left portion of the beam, between the reaction and the force F, where the shear is constant. Between the applied forces F, the moment does not change because there is no shear in this region. In the right portion of the beam, where the shear is constant and negative, the moment decreases at a steady rate from the maximum at the point of application of the force F to zero at the point of reaction.

In the truss, the bending moment is resisted by the forces in the top and bottom chords. The moment at any section should be equal to the product of the force in one chord and the distance between the two chords. We expect the chord forces in the truss to vary as the moment varies in the beam.

When we look at the distribution of forces in the top and bottom chords of the truss (Figure 3.3.1f, g), we notice two surprising features:

1. The variation of the forces in the top and bottom chords differ from one another.
2. They also differ from the distribution of the moment in the beam (there are abrupt changes, and the force distributions are not symmetrical about midspan).

In the following text, we try to understand the reason for these apparent inconsistencies.

First we look at a simple case—the chord forces in panel HJKL, which is not subjected to shear (Figure 3.3.2a). The moment to be resisted is $30F$. From this, we deduce that the force in each chord is

$$\text{Chord Force} = \frac{\text{Moment}}{\text{Truss Height}} = \frac{30F}{10} = 3F \tag{3.3.1}$$

The signs for the forces in the top and bottom chords are different. These forces must balance one another.

If we consider the internal normal stresses in the beam, we note a similar phenomenon. As shown in Figure 3.3.2b, the stress distribution may be assumed to be linear. The internal stresses may be represented by forces at the centroids of the tensile and compressive stresses (Figure 3.3.2c). The distance between the two internal forces is $2h/3$. Because we assumed h to be 15, we find each force to have a magnitude of $3F$ to balance the moment of $30F$ (Figure 3.3.1e).

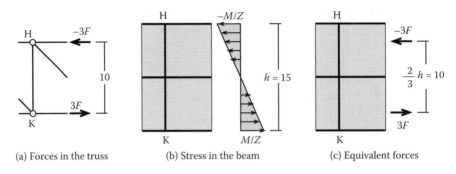

(a) Forces in the truss (b) Stress in the beam (c) Equivalent forces

FIGURE 3.3.2 Cut between H–J.

There appears to be a similarity as well as a proportionality between the internal normal stresses in beams and chord forces in trusses.

We now examine the conditions in panel ABED of the truss next to the left reaction. Our first deduction is that the vertical component of the force in the web member must be equal to F (Figure 3.3.3). Because the web member makes an angle of 45° with the horizontal, its horizontal component must also be equal to F in tension. The equilibrium of moment around node A requires zero force in the bottom chord; the equilibrium of horizontal forces requires that the top chord must carry a compressive force $-F$. That is why, in this panel, the top chord sustains a force, whereas the bottom chord does not. The moment equilibrium in that section around the top chord* gives

$$M = Fx \qquad (3.3.2)$$

which agrees with Figure 3.3.1e.

We move over to the next panel on the right (Figure 3.3.4). The shear remains the same. Therefore, the vertical and horizontal components of the web member are equal to F. Considering the equilibrium at joint B, we decide that the force in the top chord must be equal to $2F$ in compression. To maintain horizontal equilibrium

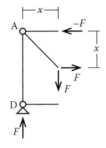

FIGURE 3.3.3 Cut between A and B.

---

* Moment equilibrium around the bottom chord also gives $M = Fx$.

Moments and Deflections in Simply Supported Beams

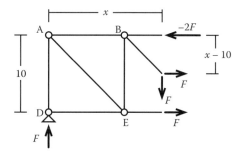

FIGURE 3.3.4 Cut between B and C.

across any section within the panel, the bottom chord force needs to be F in tension. Now we can consider the moment equilibrium in that section around the top chord:

$$M = \text{(Contribution of web member)} + \text{(Contribution of bottom chord)}$$
$$= F \times (x - 10) + F \times 10 = Fx \qquad (3.3.3)$$

This result agrees with the linearly distributed bending moment in the beam shown in Figure 3.3.1e.

From the foregoing, we deduce that (1) the abrupt changes in chord forces are caused by the condition that forces can change only at the joints of the truss and that (2) the lack of symmetry in the distribution of the chord forces is caused by the orientation of the web members. Otherwise, the distribution of the chord forces along the span of the truss represents a good analogue for the changes in the internal normal forces in a beam.

Figures 3.3.5 and 3.3.6 compare the results obtained with GOYA-A and GOYA-S, respectively. Note that the deformed shapes are also similar. The compressive and tensile forces in the highlighted region agree with the axial forces in the corresponding top and bottom chords.

Figure 3.3.7a shows the deformation of panel HJKL, where the top chord shortens and the bottom chord lengthens, as in the flexural deformation of a beam. The strains of the top and bottom chords are

$$\varepsilon = \frac{P}{EA} = \frac{3F}{EA} \qquad (3.3.4)$$

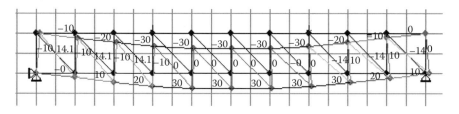

FIGURE 3.3.5 Results from GOYA-A.

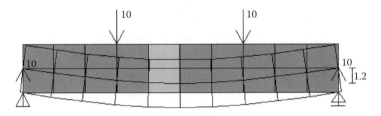

FIGURE 3.3.6 Results from GOYA-S.

where P is the axial force, E is Young's modulus, and A is the cross-sectional area of each chord. The deformation of each chord is obtained as the product of the strain ε and the length of the panel $\delta x = 10$.

$$e = \varepsilon \times \delta x = \frac{30F}{EA} \quad (3.3.5)$$

The flexural rotation $\delta\theta$ of the panel in Figure 3.3.7a (the angle between HK and JL) is

$$\delta\theta = \frac{2|e|}{10} = \frac{6F}{EA} \quad (3.3.6)$$

According to the definition in Chapter 2, the curvature ϕ is obtained by dividing $\delta\theta$ by the width $\delta x = 10$:

$$\phi = \frac{\delta\theta}{\delta x} = \frac{6F}{10EA} \quad (3.3.7)$$

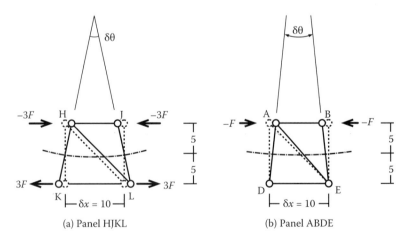

FIGURE 3.3.7 Flexural deformation.

Moments and Deflections in Simply Supported Beams

The bending moment around the centerline of the truss (the chained line in Figure 3.3.7a) is

$$M = 3F \times 5 + 3F \times 5 = 30F \tag{3.3.8}$$

Equations 3.3.7 and 3.3.8 lead to

$$\phi = \frac{M}{50EA} \tag{3.3.9}$$

On the other hand, the bending moment at the center of panel ABDE is given by substituting $x = 5$ into Equation 3.3.2:

$$M = F \times 5 = 5F \tag{3.3.10}$$

The curvature ϕ in panel ABDE is obtained in reference to Figure 3.3.7b:

$$\phi = \frac{\delta \theta}{\delta x} = \frac{F}{10EA} \tag{3.3.11}$$

Equations 3.3.10 and 3.3.11 again lead to Equation 3.3.9.

We can discuss the similarities between the truss and the beam further. We regard the truss as a beam having the section shown in Figure 3.3.8. As will be discussed in Chapter 4, the moment of inertia of the section is

$$I = 2Ay^2 \tag{3.3.12}$$

Substituting $y = 5$ into Equation 3.3.11, we get $I = 50A$. Equations 3.3.9 and 3.3.12 lead to

$$\phi = \frac{M}{EI} \tag{2.6.4'}$$

which is identical to Equation 2.6.4 we obtained in Chapter 2.

In addition to the flexural deformation shown in Figure 3.3.7b, panel ABDE is distorted (Figure 3.3.9a) because the diagonal member AE is subjected to a tensile force $\sqrt{2}F$. You can simulate the distortion if you provide GOYA-A with the boundary condition shown in Figure 3.3.9b. A similar distortion also occurs in beams. It is called *shear deformation* or *shear distortion* (note that the axial force in the diagonal member is related to the shear force in the corresponding beam).

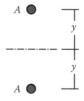

FIGURE 3.3.8 Truss section.

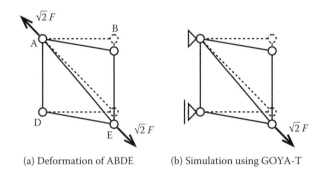

(a) Deformation of ABDE   (b) Simulation using GOYA-T

FIGURE 3.3.9 Shear deformation.

However, the effect of the shear deformation (or distortion) on the total deflection is small in shallow trusses or beams such as those shown in Figures 3.3.1 or 3.3.4.

EXERCISE

Choose any numbers i and j to set the loads shown in Figure 3.3.10. Draw axial-force diagrams similar to those in Figure 3.3.1. Draw axial-force diagrams similar to those in Figure 3.3.1. Also, draw the deformed shape. Compare your result with those obtained by a friend for the same problem.

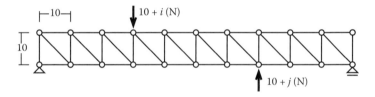

FIGURE 3.3.10 Truss subjected to two loads with opposite directions.

3.4 CONSTRUCTION AND TEST OF A TIMBER BEAM

You have made many virtual experiments with beams using GOYA. In this section, you will test an actual beam to review Chapters 2 and 3.

You need the following:

1. Two pieces of wood 1/8 in. thick and 1/4 in. wide, with length of 2 ft. In this section, we call them beams. You can buy them at a hardware shop. Do not buy balsa.
2. A ruler longer than 8 in.
3. A small plastic bag.
4. Two binder clips.
5. Two desks.
6. A kitchen scale.
7. A friend to help you.

Moments and Deflections in Simply Supported Beams

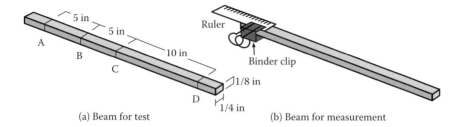

(a) Beam for test (b) Beam for measurement

FIGURE 3.4.1 Mark one of the beams and attach a ruler to the other beam.

Test 1: Measuring Young's Modulus for Wood

We shall measure Young's modulus for wood using Equation 3.1.7 developed in Section 3.1.

> **Step 1:** Mark one of the beams as shown in Figure 3.4.1a and fix the plastic bag at point C using a binder clip.
> **Step 2:** Fix the ruler to the other beam using another binder clip as shown in Figure 3.4.1b.
> **Step 3:** Lay a pencil on each desk at a distance of 20 in. as shown in Figure 3.4.2. Place the marked beam on the pencils.
> **Step 4:** Take the other beam and place the end without the ruler on the floor. Make sure it is vertical. Measure the initial location of point C.
> **Step 5:** Hang the plastic bag at midspan of the beam. Place objects (marbles, sand, etc.) in the bag so that the deflection reaches approximately 2 in. Measure the deflection at point C (v_{CC} in Figure 3.4.2) as accurately as you can.
> **Step 6:** Weigh the plastic bag with its contents using the kitchen scale.
> **Step 7:** Recall the equation for the deflection of a simply supported beam

$$v = -\frac{FL^3}{48EI}, \tag{3.1.7}$$

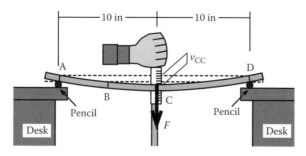

FIGURE 3.4.2 Measure deflection.

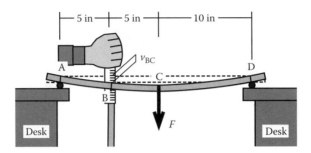

FIGURE 3.4.3 Measure deflection between support and load.

and calculate Young's modulus E by substituting the measured deflection v, the load F, the beam length L, and the moment of inertia $I = bh^3 / 12$. The unit of the load should be stated in lbf.

Step 8: Check your result noting that Young's modulus of wood (except balsa) is usually between 1000 and 1800 ksi. If your result is out of this range, reexamine your calculation and/or your measurements.

TEST 2: TESTING THE EQUATION FOR THE DEFLECTED SHAPE OF A SIMPLY SUPPORTED BEAM

In this test, we shall check Equation 3.1.5, which determines the deflected shape of the beam, using the same equipment as in Test 1.

Step 1: Calculate the deflection v_{BC} shown in Figure 3.4.3.* The notation v_{BC} indicates that the deflection is measured at point B and the force is applied at point C. Use Equation 3.1.5 and the value for E determined in Test 1.

Step 2: Measure the deflection v_{BC} and compare it with the calculation result.

TEST 3: TESTING THE RECIPROCITY THEOREM USING A SIMPLY SUPPORTED BEAM

In this test, we shall investigate the reciprocity theorem discovered by Scottish physicist James Clerk Maxwell (1831–1879), who is well known for his contributions to electromagnetism.

Step 1: Assume that the weight is at the location shown in Figure 3.4.4. Calculate the deflection at point C v_{CB} using Equation 3.1.17. This should be equal to the calculated deflection v_{BC} in Figure 3.4.3. This equality is a demonstration of Maxwell's *reciprocity theorem*.

Step 2: Do the test shown in Figure 3.4.4 to measure the deflection v_{CB}. Compare it with the calculated value.

---

* Prediction is very important in civil engineering. You need to calculated the deflection *before* the test. Car designers can test the safety of their car by testing many prototypes before actually selling the car. On the other hand, civil engineers can rarely test their design using a full-scale model; they need to predict structural behavior through calculation.

Moments and Deflections in Simply Supported Beams

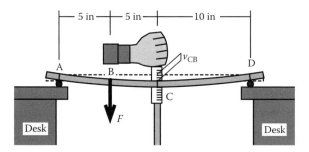

FIGURE 3.4.4 Move the load to the left string.

The reciprocity theorem is applicable to any structure. Figure 3.4.5 shows two other examples in which v_{BA} (the displacement of point B on a structure caused by a load F acting at point A) is always equal to v_{AB} (the displacement of point A caused by the same amount of load F acting at point B).

TEST 4: TESTING THE EQUATION FOR THE DEFLECTED SHAPE OF A CANTILEVER BEAM

Use the same beam and the same weight you used for Test 3.

Step 1: Place the beam at the edge of a desk and press it down as shown in Figure 3.4.6 with a stiff book or pencil case. Press it firmly. Otherwise, the beam would deform, as shown in Figure 3.4.7, and you would not obtain the correct boundary condition. (The correct boundary condition is that the beam has zero slope and zero deflection at the edge of the table or that we have a "fixed end" at C.) Measure the deflections v_{AA} and v_{BA} at locations A and B, respectively.

Step 2: Calculate the deflections v_{AA} and v_{BA} for this test using the equations in Chapter 2. Do they agree with the measurements?

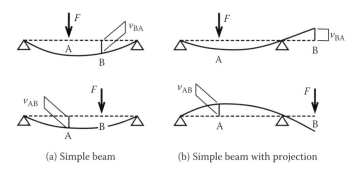

(a) Simple beam   (b) Simple beam with projection

FIGURE 3.4.5 The reciprocity theorem.

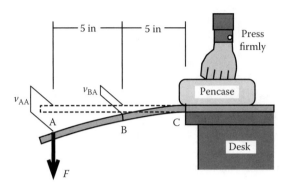

FIGURE 3.4.6 Deflections at loaded point and between support and load.

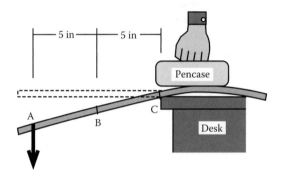

FIGURE 3.4.7 If you press the pen case insufficiently ….

TEST 5: TESTING THE RECIPROCITY THEOREM USING A CANTILEVER BEAM

Step 1: Hang the load at location B as shown in Figure 3.4.8 and measure the deflections v_{AB} and v_{BB} at locations A and B, respectively.
Step 2: Calculate the deflections v_{AB} and v_{BB} for this test using the equations in Chapter 2. Do they agree with your measurements?
Step 3: Study your results. Do they conform to the reciprocity theorem?

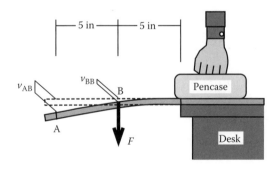

FIGURE 3.4.8 Move the load.

Moments and Deflections in Simply Supported Beams

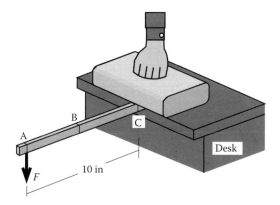

FIGURE 3.4.9 Rotate the beam around its longitudinal axis by 90°.

Test 6: Investigating the Effects of Moment of Inertia

Step 1: Rotate the beam around its longitudinal axis by 90° so that the beam has the shorter dimension of its cross section parallel to the desk surface (Figure 3.4.9).
Step 2: Measure the deflection at point A.
Step 3: Calculate the deflection and compare it with the measurement. Compare it also with the result obtained in Test 4.

Test 7: Sensing a Couple

Step 1: Hold the wood beam with your fingers at midspan.
Step 2: Ask a friend to push one end up and the other end down with the same force (Figure 3.4.10). You will sense the twist or couple that is required to resist the moment generated by equal and opposite forces applied at the ends of the beam (Figure 3.4.11a). You will also see that the deflection of the beam is antisymmetrical about midspan.
Step 3: Try holding the beam at different points away from its middle as shown in Figure 3.4.11b. Observe the deflected shape.

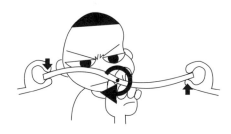

FIGURE 3.4.10 Sense a couple.

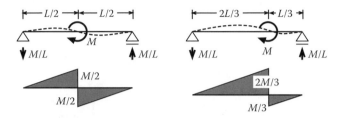

FIGURE 3.4.11 Reactions.

3.5 PROBLEMS

(Neglect self-weight in all the problems. Assume that all the beams are prismatic.)

3.1 Find an incorrect statement among the following five statements concerning the simply supported beam in Figure 3.5.1.

1. The shear force at point B is larger than that at point D.
2. The bending moment at point B is the same as that at point D.
3. The bending moment reaches maximum at point C.
4. The deflection reaches maximum at point C.
5. The slope reaches maximum at point A.

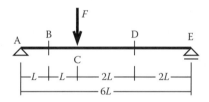

FIGURE 3.5.1 Simply supported beam with concentrated load.

3.2 Which sets of loads yield the bending-moment diagrams (1)–(5) in Figure 3.5.2? Select the correct answer from among (a)–(e) in Figure 3.5.2.

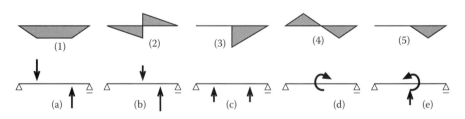

FIGURE 3.5.2 Find the correct moment distribution for each of cases (a) through (e).

Moments and Deflections in Simply Supported Beams

3.3 The width and the height of the beam in Figure 3.5.3 are 100 mm and 180 mm, respectively. What is the maximum bending stress in the beam? Select the correct answer from among (1)–(5).

(1) 50 N/mm² (2) 100 N/mm² (3) 200 N/mm²
(4) 300 N/mm² (5) 400 N/mm²

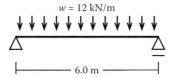

FIGURE 3.5.3 Simply supported beam loaded uniformly.

3.4 Consider a continuous beam subjected to a uniformly distributed load and supported as shown in Figure 3.5.4a. What is the ratio of the reaction at A (R_A) to that at B (R_B)? Select the correct answer from among (a)–(e). (Hint: The deflections due to a uniformly distributed load and a concentric load are shown in Figure 3.5.4b.):

(a) 1/2   (b) 1/3   (c) 2/5   (d) 3/5   (e) 3/10

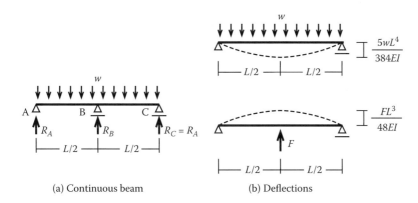

FIGURE 3.5.4 Beam continuous over two equal spans.

3.5 The beam shown in Figure 3.5.5 has zero bending moment at point A. Find the correct ratio of P to wL from among (1)–(5).

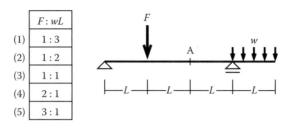

	$F:wL$
(1)	$1:3$
(2)	$1:2$
(3)	$1:1$
(4)	$2:1$
(5)	$3:1$

FIGURE 3.5.5 Beam with concentrated load and uniformly loaded overhang.

3.6 What is the deflection at point A of the beam shown in Figure 3.5.6? Select the correct answer from among (1)–(5), where I is the moment of inertia of the beam section.

(1) $\dfrac{FL^3}{8EI}$ (2) $\dfrac{FL^3}{3EI}$ (3) $\dfrac{FL^3}{2EI}$ (4) $\dfrac{2FL^3}{3EI}$ (5) $\dfrac{5FL^3}{6EI}$

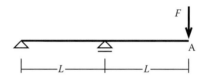

FIGURE 3.5.6 Beam with a concentrated load at the end of an overhang.

3.7 What is the deflection at point A of the beam shown in Figure 3.5.7? Select the correct answer from among (1)–(5), where I is the moment of inertia of the beam section. (Hint: use the result in Problem 3.6)

(1) $\dfrac{2FL^3}{3EI}$ (2) $\dfrac{5FL^3}{6EI}$ (3) $\dfrac{FL^3}{EI}$ (4) $\dfrac{4FL^3}{3EI}$ (5) $\dfrac{5FL^3}{3EI}$

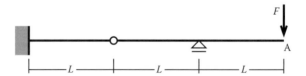

FIGURE 3.5.7 Beam with a hinge at mid-span and a concentrated load at the end of the overhanging span.

3.8 A simply supported beam is subjected to "uniformly distributed couples" as shown in Figure 3.5.8. The beam width and depth are 80 mm each, and Young's modulus is 1000 N/mm². Determine the moment diagram, the deflected shape, and the maximum deflection. (Hint: Determine

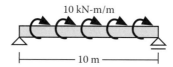

FIGURE 3.5.8 Uniformly distributed couples.

the moment diagram for the cases shown in Figure 3.5.9a, b, where the uniformly distributed couples are represented by a concentrated couple $10 \times 10 = 100$ kN-m and two concentrated couples $5 \times 10 = 50$ kN-m, respectively. Recall that these couples are represented by pairs of horizontal forces, as shown in Figure 3.5.9 c, d. (You can simulate each case using GOYA-S, as shown in Figure 3.5.9 e, f.)

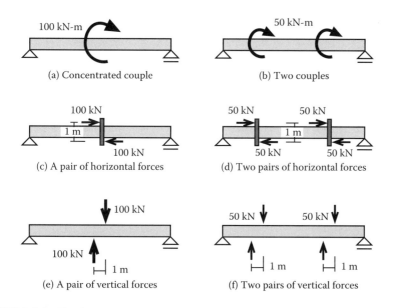

FIGURE 3.5.9 Hint for solving Problem 3.8.

4 Bending and Shear Stresses

4.1 FIRST MOMENT

In Chapters 2 and 3 we considered beams with rectangular sections and learned that

1. The stress, σ, in the extreme fiber of the section, is determined using the section modulus

$$\sigma = \frac{M}{Z}. \qquad (2.5.7)$$

2. The curvature of the beam can be determined using the moment of inertia (second moment)

$$\phi = \frac{M}{EI}. \qquad (2.6.4)$$

3. Integrating the curvature obtained by Step 2, we can determine the deflection of the beam.

These procedures apply not only for rectangular sections but also for others such as I sections and tubes. In this chapter, we shall consider sections that are not rectangular. First, we need to define the *first moment* of a section.

Start GOYA-I to reach the window in Figure 4.1.1a. Each square, surrounded by the grid lines, measures 10×10 mm. The figure shows a rectangular section with $b = 30$ mm and $h = 40$ mm. Click the "Bending stress" button to get the applet shown in Figure 4.1.1b. This applet shows the distribution of stress for a bending moment of 10×10^3 N-mm. You can change the magnitude of the moment using the sliding bar. The red and blue colors indicate compressive and tensile stresses, respectively.

Click the six squares indicated by the arrows in Figure 4.1.1a. You will get an inverted T-section, as shown in Figure 4.1.2a,b. As you click, you will notice that the red line changes its position. The red line indicates the position where the bending stress is zero, as shown in Figure 4.1.2c. It is called the *neutral axis*. To calculate the position of the axis, we shall show that the first moment * defined by the following

---

* We call this the first moment because Equation 4.1.1 includes the first order of the coordinate y. If it is replaced with the zero order of the coordinate y, i.e., $y^0 = 1$, we obtain the area of the section

$$A = \int y^0 \cdot dA = \int dA \qquad (4.1.2)$$

The *second moment* will be defined similarly in the following section.

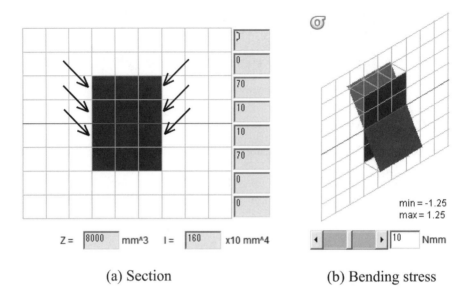

FIGURE 4.1.1 Initial windows of GOYA-I.

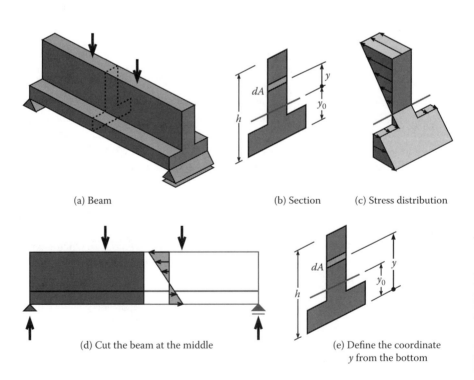

FIGURE 4.1.2 Beam of inverse-T section.

Bending and Shear Stresses

equation should be zero about the neutral axis.

$$S = \int_{-y_0}^{h-y_0} y \cdot dA = 0 \tag{4.1.1}$$

where y_0 is the distance from the bottom of the section to the neutral axis, and y is the distance from the neutral axis to an infinitesimal section dA, as shown in Figure 4.1.2b.

To derive the previous equation, let us again assume a linear strain distribution over the depth of the section as we did in Chapter 2:

$$\varepsilon = -\phi \cdot y \tag{2.5.2}$$

where ϕ is the curvature at the section. Stress at any level in the section may be expressed using Young's modulus, E, as

$$\sigma = E\varepsilon \tag{1.2.3}$$

or we may relate it to curvature using Equation 2.5.2.

$$\sigma = -E\phi \cdot y \tag{2.5.3}$$

This equation indicates that the stress is distributed linearly over the depth of the section, as shown in Figure 4.1.2c. The axial force on the section is the product of the area dA and the stress σ:

$$P = \int \sigma \cdot dA = -E\phi \int y \cdot dA = -E\phi S \tag{4.1.3}$$

The axial force in a beam subjected to transverse load is zero ($P = -E\phi S = 0$) because of the equilibrium of forces along the beam axis (Figure 4.1.2d). We therefore get $S = 0$ (Equation 4.1.1). If we redefine the y-coordinate as distance from the bottom of the section (Figure 4.1.2e), Equation 4.1.1 can be rewritten as

$$S = \int_0^h (y - y_0) \cdot dA = 0 \tag{4.1.4}$$

We can expand this equation as

$$\int_0^h (y - y_0) \cdot dA = \int_0^h y \cdot dA - y_0 \int_0^h dA = \int_0^h y \cdot dA - y_0 \cdot A = 0$$

and obtain the following equation that is more useful for calculating y_0 than Equation 4.1.1.

$$y_0 = \frac{\int_0^h y \cdot dA}{A} \tag{4.1.5}$$

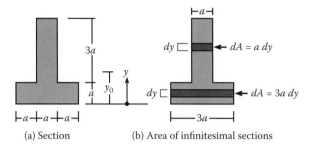

(a) Section (b) Area of infinitesimal sections

FIGURE 4.1.3 Inverted-T section.

EXAMPLE 4.1.1

Determine the location of the neutral axis y_0 for the inverse-T section shown in Figure 4.1.3a.

Solution

We shall use Equation 4.1.5 for defining the coordinate y as distance from the bottom (Figure 4.1.3). Noting that the infinitesimal area is $dA = 3a \cdot dy$ in the wide part of the section ($0 \leq y \leq a$) and $dA = a\, dy$ in the narrow part ($a \leq y \leq 3\,a$), we have

$$\int_0^h y \cdot dA = \int_0^a y \cdot 3a \cdot dy + \int_a^{4a} y \cdot a \cdot dy = 1.5a^3 + 7.5a^3 = 9a^3$$

The area of the total section is

$$A = 3a \times a + a \times 3a = 6a^2$$

Substituting this in Equation 4.1.5, we have

$$y_0 = \frac{\int_0^h y \cdot dA}{A} = \frac{9a^3}{6a^2} = 1.5a$$

EXAMPLE 4.1.2

Calculate the location of the neutral axis y_0 for the triangular section shown in Figure 4.1.4.

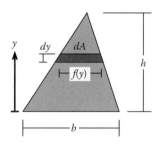

FIGURE 4.1.4 Triangular section.

Bending and Shear Stresses

Solution

We use Equation 4.1.5 again. The width of the infinitesimal area $f(y)$ shown in Figure 4.1.4 can be defined as

$$\frac{f(y)}{b} = \frac{h-y}{h}$$

Because the infinitesimal area is $dA = f(y)\,dy$,

$$\int_0^h y \cdot dA = \int_0^h y \cdot f(y)\,dy = \frac{b}{h} \cdot \int_0^h y \cdot (h-y)\,dy = \frac{bh^2}{6}$$

We get

$$y_0 = \frac{\int_0^h y \cdot dA}{A} = \frac{bh^2/6}{bh/2} = \frac{h}{3}$$

indicating that the neutral axis of a triangle crosses its centroid (i.e., center of gravity).

DETERMINING THE CENTROID OF A SECTION BY EXPERIMENT

We need the following:

1. A sheet of cardboard (preferably with a grid printed on it)
2. A thread, a needle, a pair of scissors, a ruler, and a calculator

In this test, we shall investigate if the neutral axis of any section, as in the case of a triangle, crosses its centroid.

1. Draw any section using GOYA-I.
2. Cut the cardboard to have the same shape as the section.
3. Calculate the location of the neutral axis y_0 and check to see if the result agrees with the red line in the screen. Draw the line on the cardboard model (line AB in Figure 4.1.5).

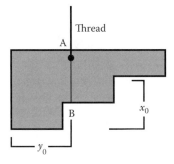

FIGURE 4.1.5 Cut-out hanging by a thread attached at A.

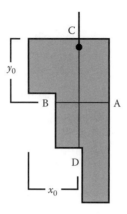

FIGURE 4.1.6 Cut-out hanging by a thread attached at C.

4. Punch a hole on line AB near the edge of the cardboard model. Hang it using the thread as shown in Figure 4.1.5. If your calculation of y_0 is correct, the line AB will be vertical. This is the proof that the neutral axis AB crosses the centroid of the board.
5. Calculate the location of the neutral axis x_0 assuming that the section in Figure 4.1.5 is bent around the horizontal axis. Hang the board as shown in Figure 4.1.6. Make certain again that the line CD is vertical. This is the proof that the neutral axis CD crosses the centroid of the board.
6. Punch another hole in the cardboard model. The hole must not be on lines AB or CD (Figure 4.1.7). Check to see if the extension of the thread (the broken line in Figure 4.1.7) crosses the intersection point of the neutral axes. The broken line represents the neutral axis when the section is bent around the broken line.

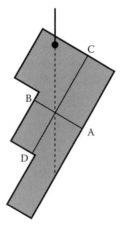

FIGURE 4.1.7 Broken line that intersects the common point of lines AB and CD represents the neutral axis for bending about the broken line.

Bending and Shear Stresses

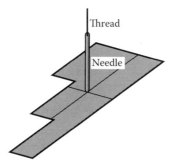

FIGURE 4.1.8 Demonstration using needle and thread, of the centroid of the shape of the board.

7. Stick the needle and thread into the cardboard model as shown in Figure 4.1.8, and make certain that the cardboard is horizontal. This is the proof that the stuck point is the centroid of the board.

Let us examine the reason why the line AB in Figure 4.1.5 is vertical. We shall use Figure 4.1.9, where the y-coordinate is measured from the neutral axis. If t is the thickness of cardboard and ρ its density, the moment around the hole caused by the slice dA is

$$dM = (\rho \cdot t \cdot dA) \times y$$

Integrating the moment over the whole body gives

$$M = \int dM = \rho \cdot t \cdot \int y \cdot dA = \rho \cdot t \cdot S$$

where S is the first moment defined by Equation 4.1.1. Because $S = 0$, we get $M = 0$, which means that the body is in equilibrium in terms of moment and does not rotate.

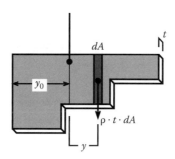

FIGURE 4.1.9 Moment around the hole.

If you conduct an integration over both the x and y directions, you can demonstrate the reason why the board is horizontal in Figure 4.1.8.

COFFEE BREAK

The first person who discovered the computation method for the center of gravity was a Greek physicist, Archimedes (BC 287–212). It is an important extension of his famous principle of the lever. Note that we also used this principle in reference to Figure 4.1.9. Archimedes also invented integral calculus, which is indispensable to the computation of the center of gravity. E. T. Bell describes him as "the greatest scientist in antiquity" in the book *Men in Mathematics**. The inventors of differential calculus, the counterpart of integral calculus, were Isaac Newton and Gottfried W. von Leibnitz, who lived in the 17th century.

4.2 SECOND MOMENT AND SECTION MODULUS

In the preceding section, we defined the *first* moment as

$$S = \int y \cdot dA \qquad (4.1.1)$$

where y is the distance between the infinitesimal section dA and the neutral axis (Figure 4.2.1). In this section, we shall define the *second* moment (or the moment of inertia) by replacing y with y^2 in Equation 4.1.1:

$$I = \int y^2 \cdot dA \qquad (4.2.1)$$

In Section 2.5, Chapter 2, we learned that the bending moment is the integral of the axial force of the infinitesimal section $\sigma \cdot dA$ multiplied by the distance from the neutral axis y:

$$M = -\int y \cdot \sigma \cdot dA \qquad (2.5.3)$$

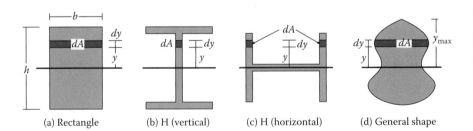

(a) Rectangle   (b) H (vertical)   (c) H (horizontal)   (d) General shape

FIGURE 4.2.1 Various sections and infinitesimal segments.

---

* Bell, Eric Temple, 1937. *Men in Mathematics*, Touchstone (Simon & Schuster), New York.

Bending and Shear Stresses

This equation applies to all kinds of sections. In Section 2.6 we learned that the stress is proportional to the curvature ϕ and the distance from the neutral axis y, as expressed in Equation 2.6.3.

$$\sigma = -E\phi y \qquad (2.6.3)$$

Substituting this into Equation 2.5.3,

$$M = E\phi \int y^2 \cdot dA = EI\phi \qquad (4.2.2)$$

The moment of inertia plays an important role in relating the bending moment to the curvature (curvature is a measure of how or at what rate the beam bends). For the rectangular section in Figure 4.2.1a, $dA = b.dy$ and

$$I = \int y^2 \cdot dA = \int_{-h/2}^{h/2} y^2 \cdot b \cdot dy = \frac{bh^3}{12} \qquad (4.2.3)$$

as we learned in Section 2.6.

Figure 4.2.2 shows the initial window of GOYA-I. As was stated earlier, each square measures 10×10 mm. The digits in the right-hand column show the contribution of each row to the moment of inertia. For example, the contribution of the uppermost row is

$$\Delta I = \int_{10}^{20} y^2 \times 30 \times dy = 70 \times 10^3 \text{ mm}^4$$

as listed in the column of numbers that appear in Figure 4.2.2. If we are interested in obtaining an approximate value, we can state the contribution of these three squares

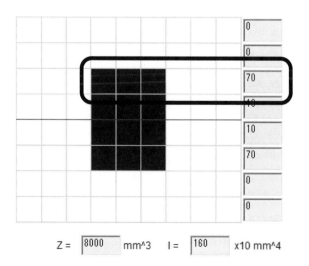

FIGURE 4.2.2 Window of GOYA-I.

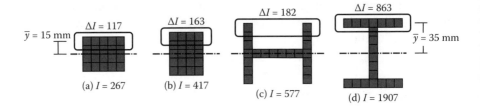

FIGURE 4.2.3 Moment of inertia (unit: 10^3 mm⁴).

to the moment of inertia as

$$\Delta I \approx \bar{y}^2 \cdot dA = 15^2 \times (30 \times 10) = 67.5 \times 10^3 \text{ mm}^4$$

where $\bar{y}$ is the distance from the neutral axis to the centroid of these squares. The total moment of inertia is shown at the bottom of the column ($I = 70 + 10 + 10 + 70 = 160 \times 10^3$ mm⁴).

Press Ctrl + N three times to create four windows. In these windows, draw the four sections shown in Figure 4.2.3. All the sections have the same area, $A = 2000$ mm², but very different moments of inertia, I, ranging from 90 to 1907×10^3 mm⁴. The large differences are caused primarily by the different contributions of the extreme rows ($\Delta I \approx \bar{y}^2 \cdot dA$ in Figure 4.2.3). For the section in Figure 4.2.3a, the average distance to the extreme rows is as small as $\bar{y} = 15$ mm, but for the section in Figure 4.2.3d the average distance is as large as $\bar{y} = 35$ mm. For the section in Figure 4.2.3c, the average distance to the extreme rows (or squares) is large but the area dA is small. The expression $\phi = M/EI$ indicates that the beam with the section in Figure 4.2.3d will have a smaller curvature and, therefore, smaller deflection for a given load over a given span than that of the other sections.

Substituting $\phi = M/EI$ into $\sigma = -E\phi y$,

$$\sigma = -\frac{y}{I} M \tag{4.2.4}$$

Press the "Bending stress" button in the windows showing the sections in Figure 4.2.3c,d and obtain the stress distributions shown in Figure 4.2.4a,b. Note that the stresses vary linearly with the distance from the neutral axis y. If we define the distance between the edge of the section and the neutral axis y_{max}, as shown in Figure 4.2.4a, the maximum stress in the section σ_{max} is

$$\sigma_{max} = \frac{y_{max}}{I} |M| \tag{4.2.5}$$

If the beam is made of brittle material with strength σ_f, it will fail at the bending moment

$$M_f = \frac{I}{y_{max}} \sigma_f \tag{4.2.6}$$

Bending and Shear Stresses

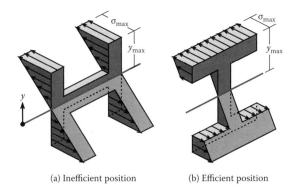

(a) Inefficient position (b) Efficient position

FIGURE 4.2.4 Stress distribution.

We call the following coefficient as the *section modulus*.

$$Z = \frac{I}{y_{max}} \tag{4.2.7}$$

In GOYA-I, Z is indicated at the bottom of the window. The section modulus of a rectangle is

$$Z = \frac{I}{y_{max}} = \frac{bh^3}{12} \bigg/ \frac{h}{2} = \frac{bh^2}{6} \tag{4.2.8}$$

as we learned in Section 2.5, Chapter 2. Now, we can rewrite Equation 4.2.6 as

$$M_f = Z\sigma_f \tag{4.2.9}$$

In other words, the strength of a beam is proportional to its section modulus. Because the section modulus of the section in Figure 4.2.4b is much larger than that of the section in Figure 4.2.4a, an I-shaped section should be positioned as shown in Figure 4.2.4b to efficiently resist bending.

Technical terms—flanges and web:

Figure 4.2.5 shows the typical section of a steel I-beam. Structural engineers call the strips in the top and bottom as *flanges*, and the vertical plate as a *web*, which

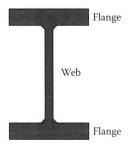

FIGURE 4.2.5 Steel I-beam.

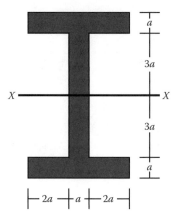

FIGURE 4.2.6 Indentification by symbols of the I-beam dimensions.

may look similar to the skin (web) that joins the toes of swans. Flanges are typically thicker than webs, as shown in the figure to resist bending moment effectively.

EXAMPLE 4.2.1

Calculate the section modulus of the section shown in Figure 4.2.6.

Solution

First, we evaluate the moment of inertia as the sum of three parts (top, middle, and bottom):

$$I = \int y^2 \cdot dA$$

$$= \int_{3a}^{4a} y^2 \cdot 5a \cdot dy + \int_{-3a}^{3a} y^2 \cdot a \cdot dy + \int_{-4a}^{-3a} y^2 \cdot 5a \cdot dy$$

$$= \frac{185}{3}a^4 + \frac{54}{3}a^4 + \frac{185}{3}a^4 = \frac{424}{3}a^4$$

Noting $y_{max} = 4a$, we have

$$Z = \frac{I}{y_{max}} = \frac{\frac{424}{3}a^4}{4a} = \frac{106a^3}{3}$$

Because the section considered is symmetrical about its neutral axis, we can shorten the calculation process by partitioning the section as shown in Figure 4.2.7—a rectangular section of $8a \times 5a$ minus two sections of $6a \times 2a$. Recalling $I = bh^3/12$ for a rectangular section, we obtain the same result.

$$I = \frac{(5a) \times (8a)^3}{12} - 2 \times \frac{(2a) \times (6a)^3}{12} = \frac{640}{3}a^4 - \frac{216}{3}a^4 = \frac{424}{3}a^4 : OK$$

Bending and Shear Stresses

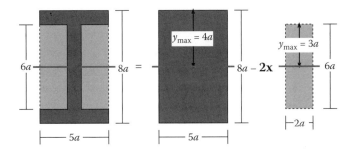

FIGURE 4.2.7 Partitioning of the section (not for section modulus).

However, we should not use this technique for calculating the section modulus because y_{max} of the outer rectangle ($4a$) is different from that of the inner ones ($3a$).

$$Z = \frac{(5a) \times (8a)^2}{6} - 2 \times \frac{(2a) \times (6a)^2}{6} = \frac{160}{3}a^3 - \frac{72}{3}a^3 = \frac{88}{3}a^3 < \frac{106}{3}a^3 : \text{NG!}$$

Also, we cannot use this shortcut for calculating the moment of inertia of a section that is not symmetrical about the horizontal axis (Figure 4.2.8) because the neutral axes of the partitioned sections are different from each other. This technique is valid only for the moment of inertia of a section symmetrical about the bending axis. The correct moment of inertia of the section in Figure 4.2.8 is

$$I = \int_{-1.5a}^{-0.5a} y^2 \cdot 3a \cdot dy + \int_{-0.5a}^{2.5a} y^2 \cdot a \cdot dy = 3.25a^4 + 5.25a^4 = 8.5a^4$$

If you use the shortcut, you will get an incorrect answer:

$$I = \frac{(3a) \times (4a)^3}{12} - 2 \times \frac{(a) \times (3a)^3}{12} = 16a^4 - 4.5a^4 = 11.5a^4 > 8.5a^4 : \text{NG!}$$

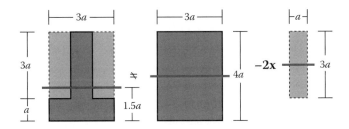

FIGURE 4.2.8 Never do this because neutral axes are different.

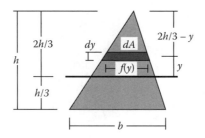

FIGURE 4.2.9 Triangular section.

Example 4.2.2

Calculate the moment of inertia and the section modulus of a triangular section (Figure 4.2.9).

Solution

Consider an infinitesimal slice of thickness dy and width $f(y)$, as shown in Figure 4.2.9. The dimension $f(y)$ can be expressed as

$$\left(\frac{2}{3}h - y\right) : f(y) = h : b \quad \text{or} \quad f(y) = \left(\frac{2}{3} - \frac{y}{h}\right) \times b$$

Because the area of the slice is $f(y)\,dy$, we have

$$I = \int y^2 \cdot dA = \int_{-h/3}^{2h/3} y^2 \cdot \left(\frac{2}{3} - \frac{y}{h}\right) \cdot b \cdot dy = \frac{bh^3}{36} \quad (4.2.10)$$

Noting that $y_{max} = 2h/3$, we obtain

$$Z = \frac{I}{y_{max}} = \frac{\frac{bh^3}{36}}{2h/3} = \frac{bh^2}{24} \quad (4.2.11)$$

Both the moment of inertia and the section modulus for the triangular section are smaller than those of the rectangular section with width b and depth h. That does not surprise us.

Example 4.2.3

Calculate the moment of inertia of a circular section with a radius of R.

Solution

We define the angle between the neutral axis and the edge of the slice, θ, as shown in Figure 4.2.10a. The width of the slice dA varies, as

$$f(y) = 2R\cos\theta$$

Figure 4.2.10b shows the segment defined by $d\theta$. Noting that $d\theta$ is so small that the arc length, $R\,d\theta$, approximates the chord length, we obtain Figure 4.2.10c, which shows in detail how we express dy in terms of $R\,d\theta$ and $\cos\theta$:

$$dy = R \cdot d\theta \cdot \cos\theta$$

Bending and Shear Stresses

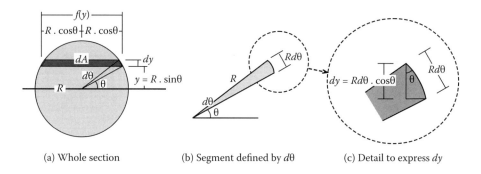

(a) Whole section (b) Segment defined by $d\theta$ (c) Detail to express dy

FIGURE 4.2.10 Circular section.

The area of the slice in Figure 4.2.10a is

$$dA = f(y) \cdot dy = 2R^2 \cdot \cos^2\theta \cdot d\theta$$

Noting that $y = R \cdot \sin\theta$ as shown in Figure 4.2.10a, the moment of inertia, I, is

$$I = \int y^2 dA = 2R^4 \int_{-\pi/2}^{\pi/2} \sin^2\theta \cos^2\theta \cdot d\theta$$

$$= \frac{R^4}{2} \int_{-\pi/2}^{\pi/2} \sin^2 2\theta \cdot d\theta = \frac{R^4}{4} \int_{-\pi/2}^{\pi/2} (1 - \cos 4\theta) \cdot d\theta = \frac{\pi R^4}{4}$$

(4.2.12)

Let us compare the preceding result with the moment of inertia of a square section having the same area, i.e.,

$$h^2 = \pi R^2$$

where h denotes the side dimension of the square. Substituting the previous equation into Equation 4.2.12,

$$I = \frac{\pi R^4}{4} = \frac{h^4}{4\pi} \approx \frac{h^4}{12.56}$$

showing that the moment of inertia of a circular section is similar to that of the square section ($I = h^4/12$) having the same area.

EXAMPLE 4.2.4

Building columns or bridge piers may be subjected to bending moment both in the x- and y-direction by earthquake or storm effects. Assume that the tube section of Figure 4.2.11a is subjected to bending moments of $M_x = M_y = 50 \times 10^6$ N-mm and compute the maximum stress in the section.* This type of column is often used in bridges.

---

* These moments are equivalent to a bending moment of $M = 50\sqrt{2} \times 10^6$ N-mm about the inclined axis in Figure 4.2.11b. See Figure 4.2.11c showing the vector summation of M_x and M_y. As we learned in Chapter 2, the bending moment itself is not a vector. However, if we cut the member and consider the forces at the cut, we can treat the moment acting on the cut as a vector. Recall that we did that for axial forces.

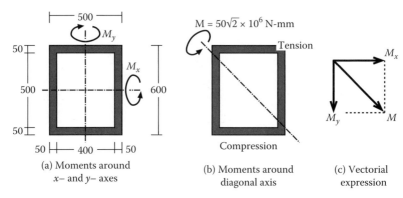

FIGURE 4.2.11 Tube section.

Solution

The moment of inertia around the x-axis is obtained by subtracting the moment of inertia of the inner rectangle (400 × 500) from that of the outer one (500 × 600):

$$I_x = \frac{500 \times 600^3}{12} - \frac{400 \times 500^3}{12} = 48.3 \times 10^8 \text{ mm}^4$$

The corresponding section modulus is

$$Z_x = \frac{I_x}{y_{max}} = \frac{48.3 \times 10^8}{300} = 16.1 \times 10^6 \text{ mm}^3$$

The maximum stress caused by the bending moment of $M_x = 50 \times 10^6$ N-mm is

$$\sigma_x = \frac{M_x}{Z_x} = \frac{50 \times 10^6}{16.1 \times 10^6} = 3.11 \text{ N/mm}^2$$

The stress distribution is shown in Figure 4.2.12a.

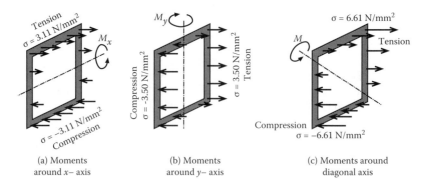

FIGURE 4.2.12 Stress distribution.

Bending and Shear Stresses

The moment of inertia around the y-axis is

$$I_y = \frac{600 \times 500^3}{12} - \frac{500 \times 400^3}{12} = 35.8 \times 10^8 \text{ mm}^4$$

I_y is smaller than I_x because of the smaller height ($h = 500$ mm). The corresponding section modulus is

$$Z_y = \frac{I_y}{y_{max}} = \frac{35.8 \times 10^8}{250} = 14.3 \times 10^6 \text{ mm}^3$$

The maximum stress caused by the bending moment of $M_y = 50 \times 10^6$ N-mm is

$$\sigma_y = \frac{M_y}{Z_y} = \frac{50 \times 10^6}{14.3 \times 10^6} = 3.50 \text{ N/mm}^2$$

The stress distribution is shown in Figure 4.2.12b.
The stress caused by the simultaneous bending moments of $M_x = M_y = 50 \times 10^6$ N-mm is shown in Figure 4.2.12c. The maximum stress is

$$\sigma = \sigma_x + \sigma_y = 3.11 + 3.50 = 6.61 \text{ N/mm}^2$$

and occurs at the corners where the moment M_x and M_y cause stresses of the same sense (tension or compression).

DETERMINE THE OPTIMUM PROPORTIONS FOR A SECTION RESISTING MOMENT

We wish to design a beam section that can resist a bending moment of $+50 \times 10^3$ N-mm (using GOYA-I) with the limitation that neither its height nor its width should exceed 80 mm. Note that the positive sign of the bending moment indicates that the bottom fiber will be in tension. Assume that we can use a material with a compressive strength of 2 N/mm² and a tensile strength of 1 N/mm². Design the section so that the area (the number of squares) is minimized. The section should be continuous, as shown in Figure 4.2.13a, and should not be discontinuous, as shown in Figure 4.2.13b. It is difficult to find the best solution directly. Do it by trial and error.

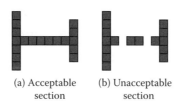

(a) Acceptable section   (b) Unacceptable section

FIGURE 4.2.13 Examples.

4.3 CONSTRUCTION AND TEST OF A STYROFOAM BEAM

We need the following:

1. A Styrofoam sheet, 2 × 2 ft, with a thickness of 1/2 in.
2. A knife
3. Adhesive for the Styrofoam
4. A weight of approximately 5 lb, such as marble or sand
5. A small plastic bag
6. A piece of wood 2 ft long
7. A ruler longer than 8 in.
8. A binder clip
9. A kitchen scale

In this section, we will make a rectangular tube beam (Figure 4.3.1) using a Styrofoam sheet. To prevent failure at the middle of the tube beam, the joint shall be strengthened, as shown in Figure 4.3.2, using the remaining material. The beam shall be located between desks placed 40 in. apart, as shown in Figure 4.3.3, and designed to fail if the load applied at midspan reaches a value of $F = 4$ lbf. The beam height (h) shall be 2 in. The beam width shall be determined based on the following assumptions.

Assumption 1: The tensile strength of Styrofoam is

$$\sigma_f \approx 30 \text{ psi}$$

Its compressive strength is higher.

Assumption 2: The beam will break at either end of the strengthened part (Figure 4.3.3), where the bending moment is

$$M = \frac{F}{2} \times L = 2 \times 18 = 36 \text{ lbf-in.}$$

Therefore, the width of the beam shall be designed to satisfy

$$M_f = Z \, \sigma_f = 36 \text{ lbf-in}$$

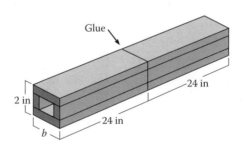

FIGURE 4.3.1 Rectangular tube beam.

Bending and Shear Stresses

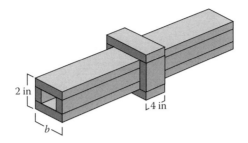

FIGURE 4.3.2 Strengthen the joint.

To make optimum use of the beam, we shall also evaluate the Young's modulus of the Styrofoam as follows:

1. Fix the ruler to the wood using a binder clip as we did in Section 3.4.
2. Hang the plastic bag on the beam.
3. Place objects (marbles, sand, etc.) into the plastic bag so that the deflection reaches approximately 2 in.
4. Measure the deflection at midspan.
5. Determine the weight.
6. Assume that the strengthened part is rigid. The deflection of the beam shall be similar to that of the cantilever beam shown in Figure 4.3.4, from which you can calculate Young's modulus.

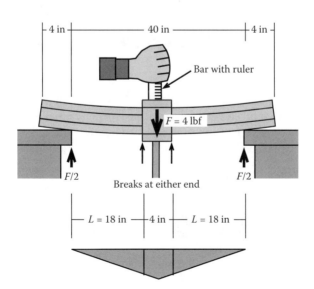

FIGURE 4.3.3 Beam under load.

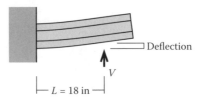

FIGURE 4.3.4 Equivalent cantilever beam.

After the calculation, increase the weight gradually until the beam fails. Then compute the strength of your Styrofoam from the test data using $\sigma_f = M_f/Z$ (i.e., assuming that Styrofoam is a brittle material).

4.4 SHEAR STRESS

We have considered tensile and compressive stresses (normal stresses) caused by bending moment. In this section, we investigate stresses caused by shear.

Figure 4.4.1a shows a body on a desk pushed by a force V. If the force is smaller than a certain threshold value, the body will not move because of the friction between

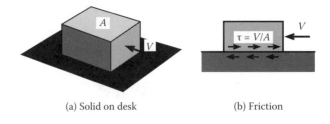

(a) Solid on desk   (b) Friction

FIGURE 4.4.1 Shear stress on desk.

Bending and Shear Stresses

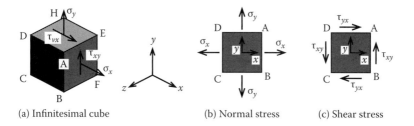

FIGURE 4.4.2 Stresses on an infinitesimal cube.

the body and the desk surface. As shown in Figure 4.4.2b, we define the average shear stress on the contact surface between the base and the desk as the ratio of the force, V, to the area of the contact surface, A:

$$\tau = \frac{V}{A} \tag{4.4.1}$$

We can define the unit shear stress on a beam section similarly by dividing the shear force on a section by the area of the section. However, the actual distribution of shear stress is not necessarily uniform in either case considered. In this section we examine shear-stress distribution on a beam section.

Before tackling the problem of shear-stress distribution over a beam section, we need to develop a set of definitions for stresses. Figure 4.4.2a shows an infinitesimal cube that is subjected to four stresses acting on the top face ADEH and side face ABEF. Axis x is perpendicular to face ABEF, and axis y is perpendicular to face ADEH.

We focus on four stresses (also see Table 4.4.1):

σ_x: the normal stress acting in the direction of the x-axis on face ABEF
σ_y: the normal stress acting in the direction of the y-axis on face ADEH
τ_{xy}: the shear stress acting in the direction of the y-axis on face ABEF
τ_{yx}: the shear stress acting in the direction of the x-axis on face ADEH

The stresses σ_x and σ_y are called normal stresses because they act in a direction perpendicular to the faces of the cube (Table 4.4.1). The stresses τ_{xy} and τ_{yx}, acting in

TABLE 4.4.1

Nomenclature for Normal and Shear Stresses

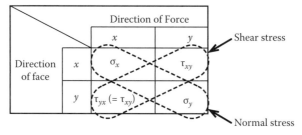

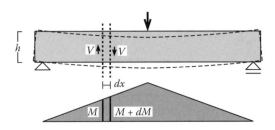

FIGURE 4.4.3 Simply supported beam and its bending-moment diagram.

directions parallel to the faces of the cube, are called shear stresses. Generally, there are normal as well as shear stresses on the face ABCD; but for simplicity we shall ignore them in order to be able to show them in two dimensions (Figure 4.4.2b,c). In Figure 4.4.2b, note that the equilibrium of forces requires a stress of σ_x on face DC and a stress of σ_y on face BC in the opposite direction to the normal stresses on faces AB and AD. In Figure 4.4.2c, similarly, τ_{xy} and τ_{yx} act on faces DC and BC. The equilibrium condition for moment leads to the following important equation.

$$\tau_{xy} = \tau_{yx} \quad (4.4.2)$$

Let us move to the problem of the beam shown in Figure 4.4.3. We assume that the beam has a rectangular section. The left and the right faces of the slice dx are subjected to bending moments M and $M + dM$, respectively. Figure 4.4.4a shows the distribution of normal stresses caused by the bending moment. Normal stresses σ_t and $\sigma_t + d\sigma_t$ denote the tensile stresses at the bottom edge of the slice. They are related to bending moments M and $M + dM$.

$$\sigma_t = \frac{M}{Z} = \frac{M}{bh^2/6} \quad (4.4.3)$$

$$\sigma_t + d\sigma_t = \frac{M + dM}{Z} = \frac{M + dM}{bh^2/6} \quad (4.4.4)$$

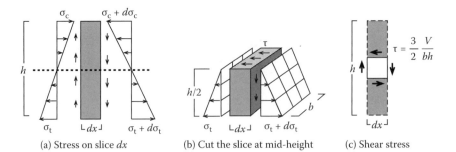

(a) Stress on slice dx (b) Cut the slice at mid-height   (c) Shear stress

FIGURE 4.4.4 Equilibrium of slice dx.

Bending and Shear Stresses

Cutting the beam slice in Figure 4.4.4a at mid-height results in the free-body diagram shown in Figure 4.4.4b. The volumes of the triangular prisms represent the tensile forces acting on the two sides of the beam slice. They are

$$T = \frac{bh}{4}\sigma_t = \frac{3}{2}\cdot\frac{M}{h} \quad (4.4.5)$$

$$T + dT = \frac{bh}{4}(\sigma_t + d\sigma_t) = \frac{3}{2}\cdot\frac{M + dM}{h} \quad (4.4.6)$$

The difference between the tensile forces, dT, is balanced by the shear force on the upper face of the free body in Figure 4.4.4b. The required force is the shear stress, τ, multiplied by the area of the face, $b.dx$.

$$\tau\cdot b\cdot dx = dT = \frac{3}{2}\cdot\frac{dM}{h} \quad (4.4.7)$$

Rearranging,

$$\tau = \frac{3}{2}\cdot\frac{dM/dx}{bh} \quad (4.4.8)$$

Recalling that $V = dM/dx$, we have the following important equation for shear stress at the neutral axis of a beam with a rectangular section:

$$\tau = \frac{3}{2}\cdot\frac{V}{bh} \quad (4.4.9)$$

Equation 4.4.9 indicates that the shear stress at the middle height of the section is 1.5 times the average shear stress, V/bh. Though the shear stress in Figure 4.4.4b is in the horizontal direction, the equilibrium of moments in the infinitesimal square, shown in Figure 4.4.4c, requires the same shear stress in the vertical direction (Equation 4.4.2). You might be surprised to find shear stresses in the horizontal direction; but if you cut the beam horizontally, as shown in Figure 4.4.5, you will see the need for glue (or shear strength) to prevent slip.

Cutting the slice dx at a distance y from mid-height, we obtain Figure 4.4.6. The volume of the solid body on the left of the slice represents the tensile force on the left face.

$$T = \frac{b}{2}\cdot\left(\sigma_t + \frac{2y}{h}\sigma_t\right)\cdot\left(\frac{h}{2} - y\right) = \frac{6M}{h^3}\left[\left(\frac{h}{2}\right)^2 - y^2\right]$$

FIGURE 4.4.5 Beam cut horizontally.

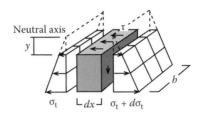

FIGURE 4.4.6 Cut the slice at y.

The tensile force on the right side, $T + dT$, can be obtained similarly. These results and $\tau = dT/(b \cdot dx)$ lead to

$$\tau = \frac{6V}{bh^3} \cdot \left[\left(\frac{h}{2}\right)^2 - y^2\right] \quad (4.4.10)$$

This equation indicates that the shear stress distribution is parabolic (Figure 4.4.7). Shear stress reaches its maximum at mid-height ($y = 0$) and is zero at the top and bottom ($y = \pm h/2$) of the section. The latter statement is intuitively obvious because the top and bottom faces of the beam are unrestrained. We cannot justify stresses acting on those surfaces. To satisfy equilibrium, the force represented by the volume of the curved solid representing the distribution of shear stress on the section (Figure 4.4.7) must be equal to the shear force in the beam.

We can similarly calculate the shear stress distribution for sections with non-rectangular shapes. Consider the section in Figure 4.4.8a. Let us compute the shear stress at distance y_1 below the neutral axis. Recall that the normal stress caused by the bending moment at position y from the neutral axis is

$$\sigma = \frac{y}{I} M \quad (4.2.4)$$

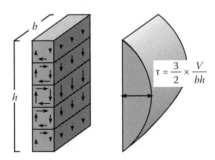

FIGURE 4.4.7 Shear stress distribution.

Bending and Shear Stresses

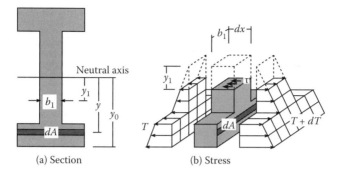

FIGURE 4.4.8 Shear stress in general section.

where I is the moment of inertia. Integrating this from the bottom (y_0) to the target (y_1), we obtain the tensile force acting on the left face of the body, as shown in Figure 4.4.8b.

$$T = \int_{y_0}^{y_1} \sigma \cdot dA = \int_{y_0}^{y_1} \frac{y}{I} M \cdot dA \qquad (4.4.11)$$

The tensile force on the right face is obtained similarly.

$$T + dT = \int_{y_0}^{y_1} (\sigma + d\sigma) \cdot dA = \int_{y_0}^{y_1} \frac{y}{I}(M + dM) \cdot dA \qquad (4.4.12)$$

From equilibrium in the horizontal direction,

$$\tau \cdot b_1 \cdot dx = dT = \int_{y_0}^{y_1} \frac{y}{I} dM \cdot dA \qquad (4.4.13)$$

where b_1 is the width of the beam at y_1. The preceding equation results in

$$\tau = \frac{V}{b_1 I} \int_{y_0}^{y_1} y \cdot dA \qquad (4.4.14)$$

Note that the integral in this equation is the first moment of the section from the bottom to y_1. Thus, we may rewrite it as follows:

$$\tau = \frac{S_1}{b_1 I} V \qquad (4.4.15)$$

where S_1 is the first moment.

EXAMPLE 4.4.1

Compute the shear stress at the neutral axis of the I-section in Figure 4.4.9a subjected to a shear force V.

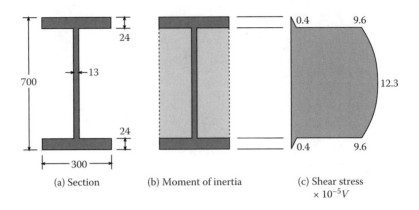

FIGURE 4.4.9 H-shaped section.

Solution

The moment of inertia of the section is obtained in accordance with Figure 4.4.9b:

$$I = \frac{1}{12} \cdot [300 \times 700^3 - (300-13) \times (700 - 2 \times 24)^3] \approx 195 \times 10^7 \text{ mm}^4$$

The first moment from the bottom to the neutral axis is

$$S = 300 \int_{-350}^{-326} y \cdot dy + 13 \int_{-326}^{0} y \cdot dy \approx 3.12 \times 10^6 \text{ mm}^3$$

Thus, the shear stress at the neutral axis is

$$\tau = \frac{S_1}{b_1 I} V = \frac{3.12 \times 10^6}{13 \times 195 \times 10^7} \approx 12.3 \times 10^{-5} V$$

We can compute the shear stresses at the other locations similarly as shown in Figure 4.4.9c. Note that the shear stress changes abruptly at the border between the flange and the web (9.6/0.4 = 24 times, in this case) because the width (b_1 in Equation 4.4.15) changes abruptly from 300 mm to 13 mm (300/13 = 24 times)*.

If we simply divide the shear force V by the area of the web, we get

$$\tau = \frac{V}{A_{web}} = \frac{V}{13 \times (700 - 2 \times 24)} \approx 11.8 \times 10^{-5} V$$

Note that this approximation is very close to the maximum shear stress shown in Figure 4.4.9c, indicating that the web resists almost all the shear force and that the contribution of the flanges to shear resistance is negligible.

---

* You may be surprised to see such an abrupt change in stress. In fact, Equation 4.4.15 is an approximate determination of the shear-stress distribution. An improved formation leads to continuous (but rapid) change at the border, though Equation 4.4.15 is sufficiently accurate for structural design.

Bending and Shear Stresses

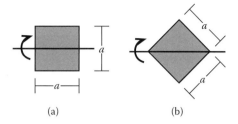

FIGURE 4.5.1 Square section.

4.5 PROBLEMS

4.1 A beam with a square section made of a brittle material is to be subjected to a moment acting about an axis parallel to an edge (Figure 4.5.1a) or about a diagonal axis (Figure 4.5.1b). About which axis is it (1) stronger and (2) stiffer?

4.2 A steel pipe has an outer radius of 500 mm, an inner radius of 480 mm, and a thickness of 20 mm (Figure 4.5.2a). Calculate the ratio of the moment of inertia of the pipe to that of the solid circular section with the same cross-sectional area. (Hint: the moment of inertia of a circular section with radius r is $I = \pi r^4/4$.) Which of the five options listed here is the correct one?

1. Approximately 5
2. Approximately 15
3. Approximately 25
4. Approximately 35
5. Approximately 45

4.3 Calculate the ratio of the moment of inertia of the I-section in Figure 4.4.9a (Section 4.4) to that of the solid square section with the same cross-sectional area. Which of the five options listed here is the correct one?

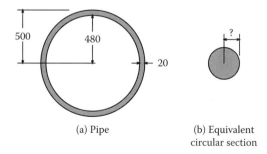

(a) Pipe

(b) Equivalent circular section

FIGURE 4.5.2 Cross sections of steel pipe and its equivalent solid section.

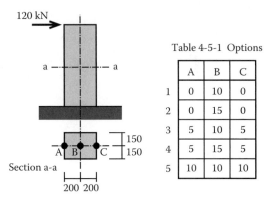

FIGURE 4.5.3 Cantilever column.

1. Approximately 5
2. Approximately 15
3. Approximately 25
4. Approximately 35
5. Approximately 45

4.4 A force of 1500 N is applied to a cantilever column with a rectangular section as shown in Figure 4.5.3. Select the correct set of shear stresses at points A, B, and C from among the options listed in Table 4.5.1. The stress unit is N/mm².

5 Frames

5.1 INTRODUCTORY CONCEPTS

The load-resisting capability of most modern buildings is provided by a structural system called a frame. It consists of beams and columns. In this section, we shall examine a beam subjected to bending moment and axial force as an introduction to understanding the response of a frame.

Start GOYA-N to reach the window shown in Figure 5.1.1. Reduce the vertical force to zero using the sliding bar in the upper right corner. You will see Figure 5.1.2a on your screen. The axial force is constant along the span ($P = F_x$) as shown in Figure 5.1.2b. The axial force causes a uniformly distributed stress in the section (Figure 5.1.2c). Recall that

$$\sigma = \frac{P}{A} \quad (A: \text{sectional area}) \tag{1.2.1}$$

Reduce the horizontal force to zero and increase the vertical load to develop the condition in Figure 5.1.3a. The shear force will be constant along the span ($P = F_x$) as shown in Figure 5.1.3b, and the bending moment is distributed linearly (Figure 5.1.3c). As discussed in Section 2.5, the normal stress on the section at the fixed end varies linearly with height of section (Figure 5.1.3d). Recall that

$$\sigma = \frac{M}{Z} \quad (Z: \text{section modulus}) \tag{2.5.7}$$

If you apply horizontal and vertical forces simultaneously (Figure 5.1.4a), the beam will elongate and deflect. The stress distribution will represent the sum or superposition of Figures 5.1.2c and 5.1.3d (Figure 5.1.4b). Note that the distribution is not symmetrical about the center of the section.

In GOYA-N, it is assumed that the beam has a width of 10 mm, a depth of 15 mm, and a length of 100 mm (these dimensions were also used in GOYA-C). The axial stress caused by the horizontal force 10 N is

$$\sigma = \frac{P}{A} = \frac{F_x}{A} = \frac{10}{10 \times 15} = 0.07 \text{ N/mm}^2$$

whereas the bending stress caused by the vertical force 10 N is

$$\sigma = \frac{M}{Z} = \frac{F_y L}{Z} = \frac{10 \times 100}{10 \times 15^2 / 6} = 2.7 \text{ N/mm}^2$$

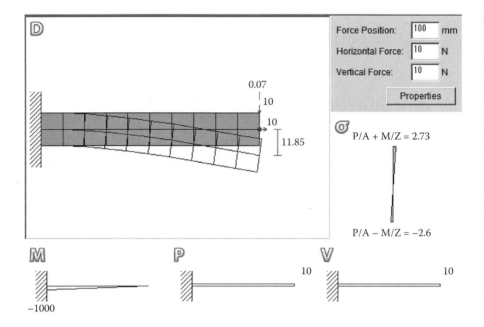

FIGURE 5.1.1 Initial window of "GOYA-N."

Note that the stress caused by the bending moment is much larger than that caused by the axial force. The asymmetry of Figure 5.1.4b is exaggerated.

How about deformations? In GOYA-N, Young's modulus is assumed to be 100 N/mm². Elongation caused by the horizontal tensile force is obtained using Equation 1.2.2.

$$u_x = \varepsilon \times L = \frac{F_x}{EA} \times L = \frac{10}{100 \times 10 \times 15} \times 100 = 0.07 \text{ mm}$$

The deflection caused by the vertical force is obtained using Equation 2.8.14.

$$u_y = \frac{F_y L^3}{3EI} = \frac{10 \times 100^3}{3 \times 100 \times (10 \times 15^3 / 12)} = 11.9 \text{ mm}$$

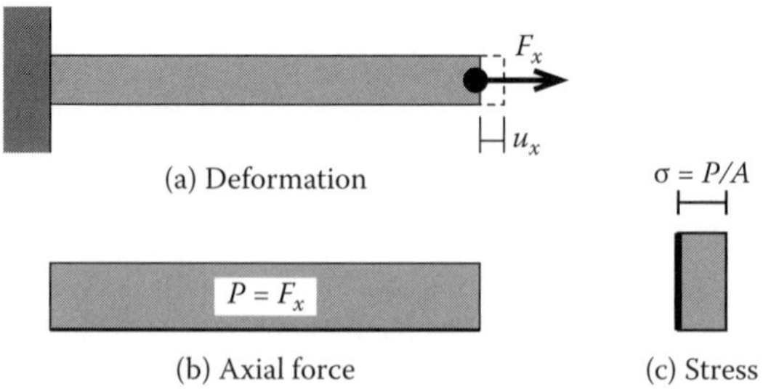

(a) Deformation

(b) Axial force $P = F_x$

(c) Stress $\sigma = P/A$

FIGURE 5.1.2 Axial force and axial stress in a beam.

Frames

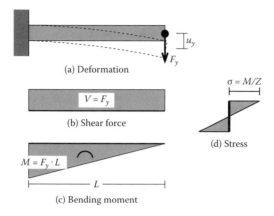

FIGURE 5.1.3 Shear force, bending moment, and bending stresses in a cantilever loaded transversely at its free end.

Note that the horizontal elongation is much smaller than the vertical deflection. Structural members are typically stiffer if loaded axially than if loaded transversely.

You can move the location of the force using the sliding bar at the upper right of the window. Move the force to midspan as shown in Figure 5.1.5a. The horizontal and vertical deformations reduce to 1/2 and 1/3, respectively. Note that the axial force, shear force, and bending moment are zero to the right of the load (Figure 5.1.6).

EXAMPLE 5.1.1

Construct the axial-force, shear-force, and bending-moment diagrams for the beam shown in Figure 5.1.7.

Solution

We can decompose the applied forces into horizontal and vertical components as shown in Figure 5.1.8 to obtain the axial-force, shear-force, and bending-moment diagrams in Figure 5.1.9.

EXAMPLE 5.1.2

Construct the axial-force, shear-force, and bending-moment diagrams for the beam shown in Figure 5.1.10.

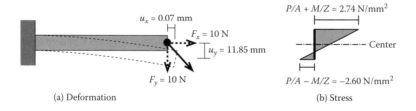

FIGURE 5.1.4 Deflected shape and bending stress distribution in a cantilever with a force no collinear with beam axis.

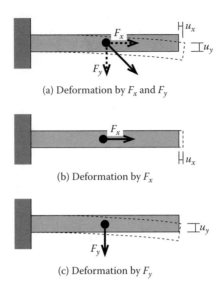

(a) Deformation by F_x and F_y

(b) Deformation by F_x

(c) Deformation by F_y

FIGURE 5.1.5 Move the force to mid-span.

Solution

We can decompose the applied forces into their horizontal and vertical components (Figure 5.1.11a,b) to obtain the axial-force, shear-force, and bending-moment diagrams in Figure 5.1.11c–e. The axial forces are shown as negative because they are compressive.

EXAMPLE 5.1.3

Construct the axial-force, shear-force, and bending-moment diagrams for the beam shown in Figure 5.1.12.

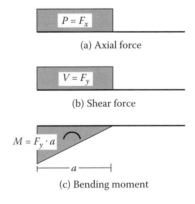

(a) Axial force

(b) Shear force

(c) Bending moment

FIGURE 5.1.6 Internal forces.

Frames

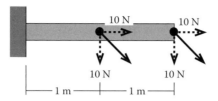

FIGURE 5.1.7 Beam under two forces.

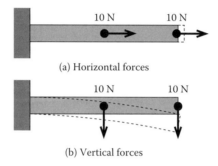

FIGURE 5.1.8 Decomposed forces.

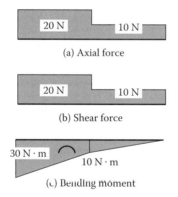

FIGURE 5.1.9 Internal forces.

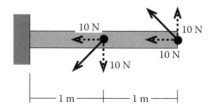

FIGURE 5.1.10 Beam under two forces.

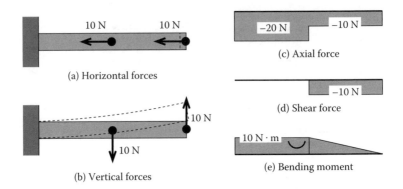

FIGURE 5.1.11 Solution.

Solution

The horizontal and vertical components of the applied force induce the reactions shown in Figure 5.1.13a,b, leading to the axial-force, shear-force, and bending-moment diagrams in Figure 5.1.13c–e. Note that the axial force is zero in the right half of the beam.

EXAMPLE 5.1.4

Construct the axial-force, shear-force, and bending-moment diagrams for the beam shown in Figure 5.1.14.

Solution

The reactions are shown in Figure 5.1.15a. Please note that there are no horizontal components because the roller support cannot resist any horizontal force. Figures 5.1.15b,c show the axial and transverse components of the applied force and the reactions. Thus, we get the axial-force, shear-force, and bending-moment diagrams for the beam shown in Figure 5.1.15d–f. This example is similar to the truss of Problem 1.12 in Section 1.11, Chapter 1.

EXAMPLE 5.1.5

Construct the axial-force, shear-force, and bending-moment diagrams for the beam shown in Figure 5.1.16.

Solution

Let us denote the reactions as shown in Figure 5.1.17. Note that there is no vertical reaction at roller support A.

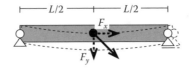

FIGURE 5.1.12 Simply supported beam.

Frames 233

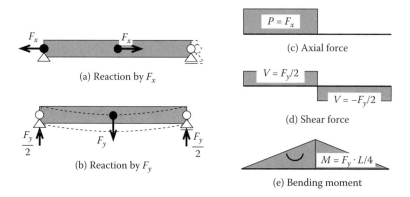

FIGURE 5.1.13 Solution.

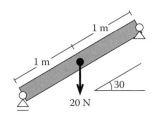

FIGURE 5.1.14 Inclined beam.

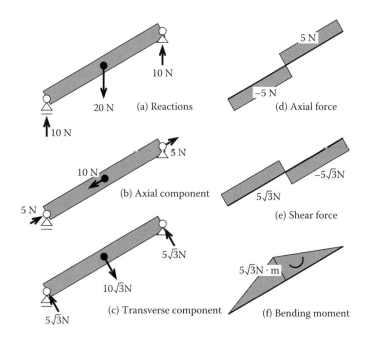

FIGURE 5.1.15 Internal forces.

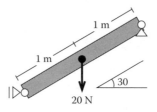

FIGURE 5.1.16 Inclined beam supported by vertical roller.

Force equilibrium, $\Sigma X = 0$ and $\Sigma Y = 0$, leads to the following equations, where the plus sign represents the force to the right or up.

$$R_{Ax} + R_{Bx} = 0$$

$$-20 + R_{By} = 0$$

Moment equilibrium, $\Sigma M = 0$, around point B, leads to the following equation, where the minus sign represents counter-clockwise moment.

$$-20 \times \frac{\sqrt{3}}{2} - R_{Ax} \times 1 = 0$$

Solving these equations, we get the reactions shown in Figure 5.1.18a. Figure 5.1.18b,c show the axial and transverse components of the applied force and the reactions, leading to the axial-force, shear-force, and bending-moment diagrams for the beam shown in Figure 5.1.18d–f. This example is similar to the truss of Problem 1.13 in Section 1.11, Chapter 1. Imagine how the beam deforms: Point A moves down and the beam elongates. This causes the tensile axial force shown in Figure 5.1.18d.

EXAMPLE 5.1.6

Three forces are applied at the top of a cantilever column as shown in Figure 5.1.19. Calculate the normal stress distribution at the fixed end.

Hint: Evaluate the stresses caused by the axial force (Figure 5.1.20a) and the bending moments around the x- and y-axes (Figure 5.1.20b,c).

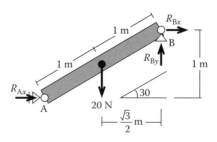

FIGURE 5.1.17 Reactions.

Frames 235

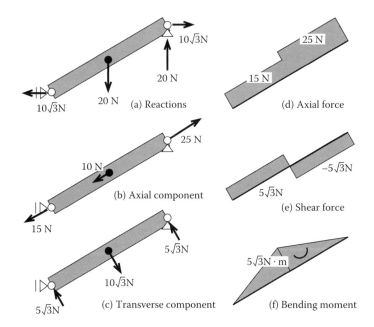

FIGURE 5.1.18 Internal forces.

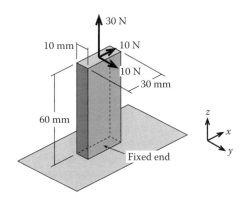

FIGURE 5.1.19 Cantilever column.

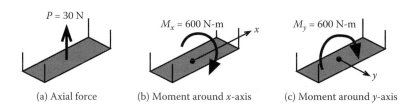

FIGURE 5.1.20 Hint.

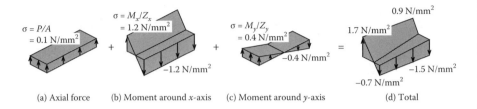

（a) Axial force　(b) Moment around x-axis　(c) Moment around y-axis　(d) Total

FIGURE 5.1.21 Solution.

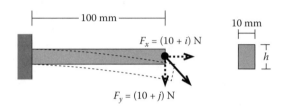

FIGURE 5.1.22 Determine the depth of a cantilever beam subjected to a specified load.

Solution

The stresses caused by the axial force and the bending moments are shown in Figure 5.1.21a–c, where Z_x and Z_y represent the section moduli with respect to x- and y-axes:

$$Z_x = \frac{30 \times 10^2}{6} = 500 \text{ mm}^3 \quad \text{and} \quad Z_y = \frac{10 \times 30^2}{6} = 1500 \text{ mm}^3$$

Superposing these results leads to the solution shown in Figure 5.1.21d.

Design Your Beam

Using single digits i and j of your choice, define the forces shown in Figure 5.1.22. The material used for the beam has tensile and compressive strengths of 3 and 2 N/mm², respectively. The beam width is 10 mm. Determine the required beam depth ignoring the effect of shear stress.

5.2 A SIMPLE BENT

Start GOYA-L to find the "simple bent" shown in Figure 5.2.1. Vary the vertical force F_y using the sliding bar in the lower left corner. You will see that the free end C displaces not only vertically but also horizontally whether the force F_y acts up or down. Looking at the deflected shape, we can understand this intuitively. The horizontal member AB bends.* This results in deflection and slope changes at corner B. The vertical member BC does not bend, but the vertical position of the free end C

---

* The horizontal displacement, u_x, is assumed to be small enough compared with the length, L.

Frames

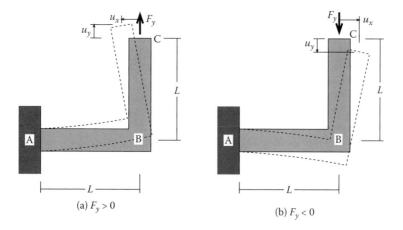

FIGURE 5.2.1 A simple bent

changes because of the deflection at B, and its horizontal position changes as a result of the slope at B.

In this section, we shall learn how to determine the deflected shape of the bent analytically.

Cut the vertical member between B and C (Figure 5.2.2a). You will find that a tensile axial force $P = F_y$ acts at the cut to maintain equilibrium. Now, cut the bent at the horizontal member between A and B (Figure 5.2.2b). You will find a negative (counter-clockwise) shear force $V = -F_y$ and a positive (concave upward) bending moment $M = F_y(L - x)$. Distributions of axial force, shear force, and bending moment are plotted in Figure 5.2.3.

From the moment diagram in Figure 5.2.3c, we understand that the horizontal member deforms like a cantilever beam (Figure 5.2.4a). The slope and deflection of

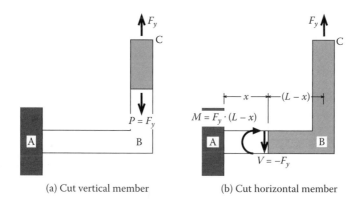

FIGURE 5.2.2 Free body diagram.

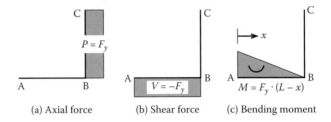

FIGURE 5.2.3 Internal forces.

point B are obtained using the equations for a cantilever beam subjected to a force applied at the free end (Section 2.8, Chapter 2).

$$\theta = \frac{F_y L^2}{2EI} \quad (2.8.12)$$

$$v = \frac{F_y L^3}{3EI} \quad (2.8.14)$$

Note that the vertical member remains straight (Figure 5.2.4b) because there is no bending moment acting on it. Also note that member BC rotates by θ because it needs to remain perpendicular to member AB at point B. We conclude that the horizontal displacement of end C is

$$u_x = -\theta \times L = -\frac{F_y L^3}{2EI} \quad (5.2.1)$$

where the minus sign indicates that the top end displaces to the left. How about the vertical displacement? Because the axial deformation of the vertical member caused by the axial force $P = F_y$ is negligible in comparison with the upward deflection of

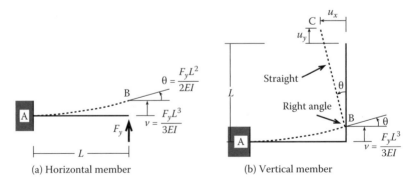

FIGURE 5.2.4 Deformation.

Frames

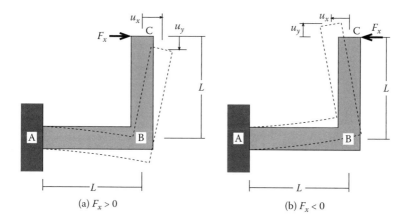

FIGURE 5.2.5 Horizontal force.

member AB (Section 5.1), we conclude that the vertical displacement of end C is equal to the deflection of the horizontal member at B.

$$u_y = v = \frac{F_y L^3}{3EI} \tag{5.2.2}$$

Reduce the vertical force to zero and apply a horizontal force to the right (Figure 5.2.5a) or to the left (Figure 5.2.5b). You see that the horizontal and vertical displacements are much larger than they were when the vertical force was applied. Why?

Cut the vertical member between B and C (Figure 5.2.6a). You will find a shear force $V = F_x$ and a bending moment $M = F_x(L - y)$ at the cut (coordinate y defines the height of the cut). If you cut the horizontal member (Figure 5.2.6b), you will find a tensile axial force $P = F_x$ and a bending moment* $M = F_x L$. Distributions of axial force,

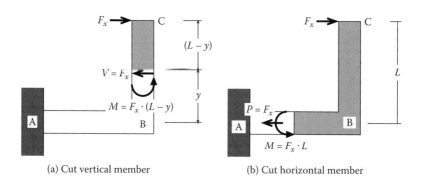

FIGURE 5.2.6 Free body diagram.

---

* According to the definition in Chapter 2, the bending moment is negative because it is concave downward. However, this sign system is confusing when dealing with sophisticated structures. From here, we shall ignore the sign of the bending moment. As long as you know which side is compressed, you can ignore the sign.

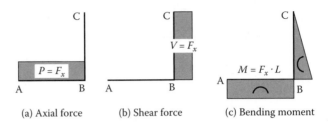

FIGURE 5.2.7 Internal forces.

the shear force, and bending moment are plotted in Figure 5.2.7. The moment does not change with distance from point A (Figure 5.2.7c) because there is no shear in the horizontal member (Figure 5.2.7b). Note that the bending moment is *continuous* at point B (Figure 5.2.7c): The moments at point B acting on the horizontal member and vertical member are the same. The joint between the two members, considered ideally to be a point in this case, has to be in equilibrium. This is an important feature of frames with continuous members.

From the sense (sign) of the moment diagram in Figure 5.2.7c, we understand that the horizontal member deforms as shown in Figure 5.2.8a. The slope and deflection at point B are obtained using the equations for a cantilever beam subjected to uniform bending moment $M = F_x L$ (Section 2.8, Chapter 2).

$$\theta = \frac{F_x L^2}{EI} \tag{2.8.9}$$

$$v = -\frac{F_x L^3}{2EI} \tag{2.8.10}$$

The vertical member does not remain straight as in the previous case. It bends (Figure 5.2.8b). Also note that the vertical member rotates by θ at its base because it is connected to remain perpendicular to the horizontal member at point B.

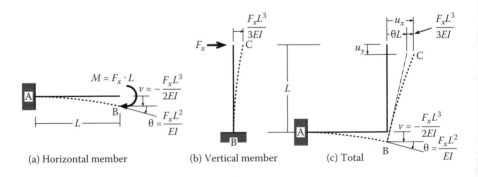

FIGURE 5.2.8 Deformation.

Frames

We conclude that the horizontal displacement of the top end is the sum of the deflections related to base rotation and bending of the vertical member:

$$u_x = \theta L + \frac{F_x L^3}{3EI} = \frac{4F_x L^3}{3EI} \tag{5.2.3}$$

The vertical displacement of the top end is the same as that at end B of the horizontal member.

$$u_y = v = -\frac{F_x L^3}{2EI} \tag{5.2.4}$$

Compare these results with Equation 5.2.1 and 5.2.2; the horizontal and vertical displacements caused by F_x are 2.7 and 1.5 times those caused by $F_y (= F_x)$.

FOR INTERESTED READERS: JOINT

An interested reader may ask the following questions:

1. In the previous discussion, we assumed that the lengths of the horizontal and vertical members are L. The actual lengths are, however, $L - h/2$ as shown in Figure 5.2.9a. If we recognize that, would the displacement be determined to be smaller than the results discussed?
2. Does recognition of member thickness affect the axial-force, shear-force, and bending-moment diagrams?
3. What happens within the joint enclosed by the broken lines in Figure 5.2.9a? Does it deform?

The answer to the first question is yes. But the effect is small if L is much larger than h. The answer to the second question is no. You can check it by cutting the bent at any place: h does not affect equilibrium. The third question is not so simple.

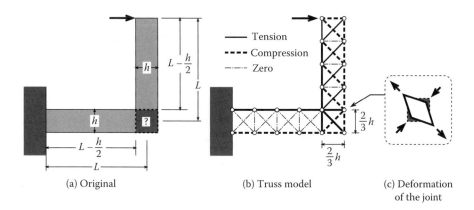

(a) Original   (b) Truss model   (c) Deformation of the joint

FIGURE 5.2.9 Beam-column joint.

242 Understanding Structures: An Introduction to Structural Analysis

To answer it, we may analyze a truss model as shown in Figure 5.2.9b, where the solid and broken lines indicate tensile and compressive members, respectively. Note that the joint is stressed diagonally so that the joint deforms as shown in Figure 5.2.9c. This deformation increases the displacement at the top end of the bent, but not significantly in most cases. When you design an actual bent, however, you should pay attention to the effects of stresses in the joint to make it safe.

EXAMPLE 5.2.1

Construct the axial-force, shear-force, and bending-moment diagrams for the bent in Figure 5.2.10. Also, calculate the displacement at point C.

Solution

Cut members AB and BC as shown in Figure 5.2.11 to obtain the axial-force, shear-force, and moment diagrams (Figure 5.2.12a–c). We can also obtain Figure 5.2.12c by superposing the moment diagrams related to the horizontal (F) and vertical ($2F$) forces illustrated in Figure 5.2.12d. The moment diagram in Figure 5.2.12c is the algebraic sum of those in Figure 5.2.12d.

The displacement can be obtained by either one of two methods. One is to use Equations 5.2.1 through 5.2.4, substituting $F_y = -2F$ and $F_x = -F$. Another is to use the following equation for the deflection of a cantilever beam:

$$v = \frac{FL^3}{3EI} \tag{2.8.14}$$

The deflection at midspan of the horizontal member is obtained substituting $-2F$ for F and $L/2$ for L because the bending moment is zero at midspan (Figure 5.2.13a):

$$v_m = -\frac{2F}{3EI}\left(\frac{L}{2}\right)^3 = -\frac{FL^3}{12EI} \tag{2.8.14''}$$

The deflected shape of the horizontal member is symmetrical about its midspan; thus, the vertical displacement of the top end is $u_y = -2v_m$ (Figure 5.2.13b). Note that the

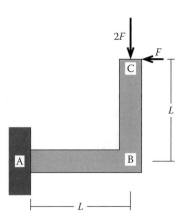

FIGURE 5.2.10 Bent with two forces.

Frames

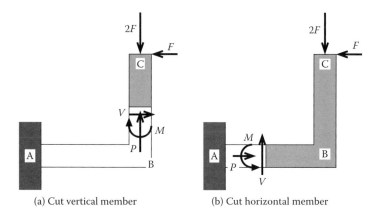

(a) Cut vertical member (b) Cut horizontal member

FIGURE 5.2.11 Free body diagrams.

slope of the horizontal member is zero at B. Therefore, the horizontal displacement u_x is obtained using Equation 2.8.14 again as shown in Figure 5.2.13b.

EXAMPLE 5.2.2

Construct the axial-force, shear-force, and bending-moment diagrams for the bent in Figure 5.2.14. Also calculate the displacement at point C.

Solution

Considering the free-body diagrams of the horizontal and vertical members as we did in Figure 5.2.11, we get the diagrams shown in Figure 5.2.15a–c. We can also obtain Figure 5.2.15c by superposing the effects of the vertical and horizontal forces as we did in Figure 5.2.12d.

The horizontal member deforms in the same way as in the previous example because the bending moment is the same. Therefore, the vertical displacement u_y is equal to that in the previous example, as shown in Figure 5.2.16c. To evaluate the deformation of the vertical member, imagine a cantilever beam (Figure 5.2.16a). Because the member is straight where the bending moment is zero (Figure 5.2.16b), the deflection of the free end is

$$v = \theta \cdot \frac{L}{2} + v_m = -\frac{5FL^3}{24EI} \qquad (2.6.25')$$

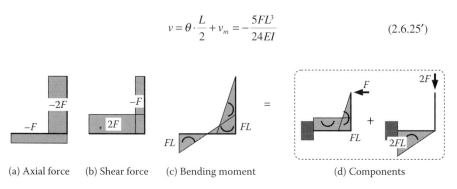

(a) Axial force (b) Shear force (c) Bending moment (d) Components

FIGURE 5.2.12 Internal forces.

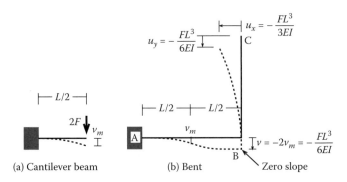

FIGURE 5.2.13 Deformation.

as we found in Example 2.6.4, Chapter 2. Recall that the slope at point B is zero (Figure 5.2.16c). Therefore, the horizontal displacement u_x is equal to the deflection of the cantilever in Figure 5.2.16b.

EXAMPLE 5.2.3

Construct the axial-force, shear-force, and bending-moment diagrams for the bent in Figure 5.2.17. Also, calculate the displacement at point E.

Solution

Consider the free-body diagrams shown Figure 5.2.18. You will note that the shear force is positive (clockwise) in Figure 5.2.18a, zero in Figure 5.2.18b, and negative (counter-clockwise) in Figure 5.2.18c. This leads to the diagrams shown in Figure 5.2.19.

To calculate the displacement, we use the deflection of a cantilever beam in Figure 5.2.20a, again. Because the moment diagram is symmetric about midspan (point B), the displacement of point C is twice the deflection of point B.

$$v = \frac{2FL^3}{3EI} \qquad (5.2.5)$$

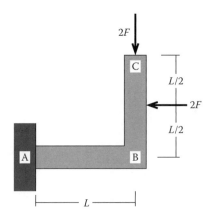

FIGURE 5.2.14 Bent with two forces.

Frames

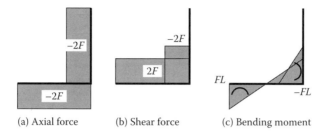

(a) Axial force   (b) Shear force   (c) Bending moment

FIGURE 5.2.15 Internal forces.

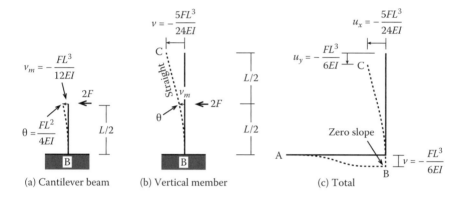

(a) Cantilever beam   (b) Vertical member   (c) Total

FIGURE 5.2.16 Deformation.

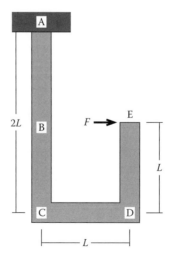

FIGURE 5.2.17 Bent with two joint.

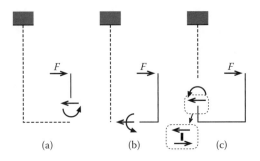

FIGURE 5.2.18 Free body diagrams.

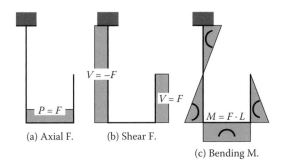

FIGURE 5.2.19 Internal forces.

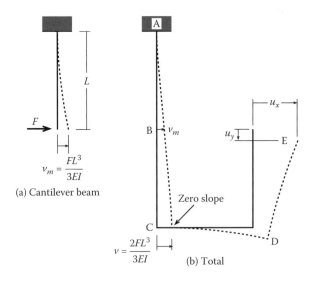

FIGURE 5.2.20 Deformation.

Frames

Noting that the slope at point C is zero, we can conclude that the displacement of the loaded end is obtained as the displacement of point C plus the result obtained in Figure 5.2.8.

$$u_x = \frac{2FL^3}{3EI} + \frac{4FL^3}{3EI} = \frac{2FL^3}{EI} \qquad (5.2.6)$$

$$u_y = -\frac{FL^3}{2EI} \qquad (5.2.7)$$

EXAMPLE 5.2.4

Construct the bending-moment diagram of the structure in Figure 5.2.21a. Select the correct deformed shape among those in Figure 5.2.21b–d. (Hint: look at the deformation of the supporting beam, AB.)

Solution

The reaction at support A (R_A in Figure 5.2.22a) is obtained by moment equilibrium around support B ($R_A \times L - F \times c = 0$):

$$R_A = \frac{c}{L} F \qquad (5.2.8)$$

The reaction at support B (R_B in Figure 5.2.22a) is obtained by moment equilibrium around support A:

$$R_B = \frac{a+b}{L} F \qquad (5.2.9)$$

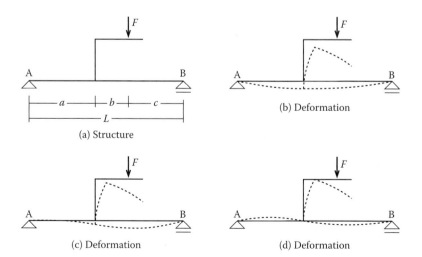

(a) Structure

(b) Deformation

(c) Deformation

(d) Deformation

FIGURE 5.2.21 Which deflected shape is correct?

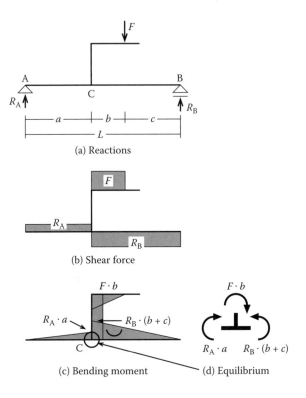

FIGURE 5.2.22 Solution.

The shear-force diagram is obtained as shown in Figure 5.2.22c. This leads to the bending-moment diagram in Figure 5.2.22c. At joint C, bending moments balance as shown in Figure 5.2.22d or

$$Fb + R_A a - R_B(b+c) = 0 \qquad (5.2.10)$$

You can confirm Equation 5.2.10 if you substitute Equations 5.2.8 and 5.2.9 into Equation 5.2.10:

$$Fb + \left(\frac{c}{L}F\right)a - \left(\frac{a+b}{L}F\right)(b+c) = 0 \qquad (5.2.11)$$

Note that the bending moment in beam AB is concave upward everywhere. Therefore, we conclude that Figure 5.2.21b is correct. You should check the results using GOYA-D.

EXERCISE

Take two numbers i and j to determine the location of the vertical member in Figure 5.2.23. Draw the bending-moment diagram and sketch the deformations.

Frames

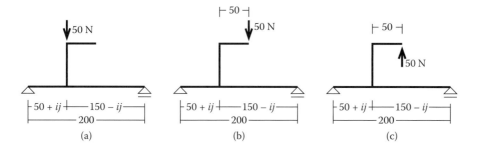

FIGURE 5.2.23 Structure.

EXAMPLE 5.2.4

Figure 5.2.24 shows a bent supported by a vertical roller at point C. Assume that the bent is subjected to a vertical force F_y. Determine the reaction R, the bending-moment diagram, and the deflection of point C.

Hint: Use the superposition technique in Example 2.8.4, Chapter 2, for a beam with a fixed end and a roller support.

Solution

Recall that, in Figure 5.2.1b, the downward force on a simple bent ($F_y < 0$) causes movement of point C to the right (Equation 5.2.1). On the other hand, a force to the left (Figure 5.2.5b) causes movement of point C to the left (Equation 5.2.3). To make the lateral movement zero, we set the right-hand terms of the two equations equal to each other.

$$\frac{F_y L^3}{2EI} = \frac{4 F_x L^3}{3EI} \qquad (5.2.12)$$

In this case, the horizontal force F_x is equivalent to the reaction R. Thus, we obtain

$$R = F_x = \frac{3}{8} F_y \qquad (5.2.13)$$

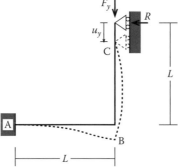

FIGURE 5.2.24 Deflected shape of bent.

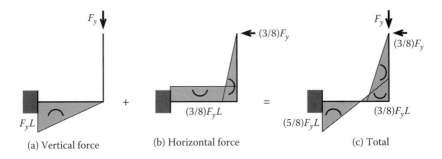

(a) Vertical force   (b) Horizontal force   (c) Total

FIGURE 5.2.25 Bending-moment diagrams.

The bending-moment diagram is obtained by superposing those given by F_y, and $F_x = (3/8)F_y$, as shown in Figure 5.2.25. The deflection is obtained similarly superposing the values given by Equation 5.2.2 and 3/8 times that given by Equation 5.2.4 (see Figure 5.2.26).

$$u_y = \frac{F_y L^3}{3EI} - \frac{(3/8)F_y L^3}{2EI} = \frac{7F_y L^3}{48EI} \qquad (5.2.14)$$

Note that the deflection is less than one half of that given by F_y. In other words, the vertical roller makes the structure stiff. Check this conclusion using GOYA-L.

DESIGN YOUR STRUCTURE

Assume any one-digit number i. We want to design a structure shown in Figure 5.2.27 that can carry a force $F = 50$ N. Assume that the material strength is 3 N/mm² and Young's modulus is 100 N/mm². You can neglect its self-weight. The cross section of the member shall be square. What is the required cross section? How much is the deflection?

Hint: Reexamine the process leading to Equations 5.2.1 and 5.2.3 very carefully and derive equations that are applicable to a bent in which the length of horizontal member is different from that of the vertical member. Then, follow the process in Example 5.2.4. Use $M = Z \cdot \sigma$ to determine the cross section. Check your result using GOYA-L.

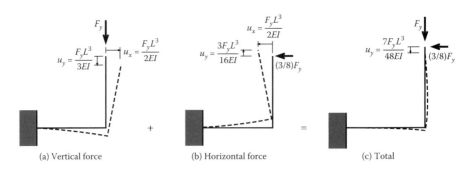

(a) Vertical force   (b) Horizontal force   (c) Total

FIGURE 5.2.26 Deflected shapes corresponding to bending-moment diagrams in Figure 5.2.25.

Frames

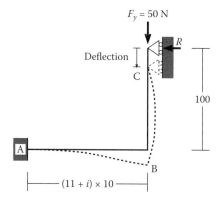

FIGURE 5.2.27 Deflected shape of bent subjected to a vertical load.

5.3 A PORTAL FRAME

Start "GOYA-P" to find the portal frame* shown in Figure 5.3.1. Apply a horizontal force at the support D. You will find diagrams showing distributions of axial force, shear force, and bending moment (Figure 5.3.2).

Figure 5.3.3a shows possible reactions for the frame as well as the applied force. The equilibrium of horizontal forces requires $R_{Ax} = F_D$. The equilibrium of vertical forces and moments requires $R_{Ay} = R_{Dy} = 0$. The reactions should be as shown in Figure 5.3.3b.

As we have done in the previous section, we consider the free-body diagrams shown in Figure 5.3.4 to obtain the axial-force, shear-force, and bending-moment diagrams shown in Figure 5.3.2. Note that the free-body diagrams in Figure 5.3.4b,c yield axial forces and bending moments having the same magnitudes.

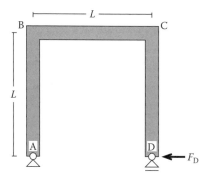

FIGURE 5.3.1 Portal frame with horizontal force at support.

---

* A frame is a structural system that may include many beams and columns. In this section, we consider a "portal frame" that includes only three members: two columns and a beam.

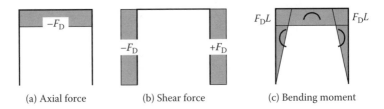

FIGURE 5.3.2 Internal forces.

Figure 5.3.5 shows the deflected shape of the portal frame. Make a special effort to relate the deflected shape of the frame to its bending-moment diagram (Figure 5.3.2c). Note that the moment causes tension on the outside and compression on the inside for all three members of the frame. That condition suggests that all three members should be bending "concave out." The deflected shape confirms this expectation. To become a competent structural engineer, one should always go through this type of reasoning and relate bending-moment distribution to deflected shape.

Because the bending moment in the beam is uniform and the slope at midspan is known to be zero in this symmetric case, the slope at beam end (θ in Figure 5.3.5b) is determined to be

$$\theta = \int_0^{L/2} \frac{M}{EI} dx = \frac{ML}{2EI} = \frac{F_D L^2}{2EI} \tag{5.3.1}$$

The displacement of joint B is obtained by superposing the contributions of the slope of the beam ($\theta.L$) and the flexural deformation of column AB given by Equation 2.8.14, Chapter 2.

$$u_B = \theta L + \frac{F_D L^3}{3EI} = \frac{5F_D L^3}{6EI} \tag{5.3.2}$$

Note that the deformation is symmetrical: i.e., column CD deforms in the same way as column AB. Therefore, the displacement of support D is

$$u_x = 2u_B = \frac{5F_D L^3}{3EI} \tag{5.3.3}$$

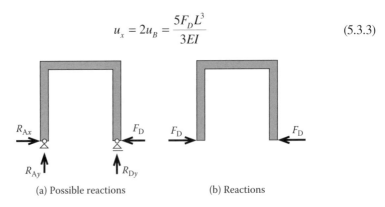

(a) Possible reactions (b) Reactions

FIGURE 5.3.3 Reactions.

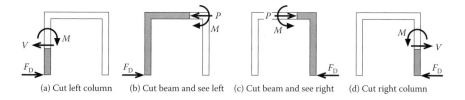

(a) Cut left column   (b) Cut beam and see left   (c) Cut beam and see right   (d) Cut right column

FIGURE 5.3.4 Cut the portal frame.

The length L is assumed to be 100 mm in GOYA-P. Change the height of the member section to 22.9 mm so that EI will be 1×10^6 N/mm² and apply a force of $F = 6$ N. You will obtain $u_x = 5 \times 6 \times 10^6 / (3 \times 10^6) = 10$ mm.

Assume a vertical load on the portal frame as shown in Figure 5.3.6a. Select the plausible deflected shape among those shown in Figure 5.3.6b–e.

To solve the problem, we need to determine the reactions. See Figure 5.3.7. Equilibrium of horizontal forces yields $R_{Ax} = 0$. Equilibrium of vertical forces and moments requires $R_{Ay} = R_{By} = F_y / 2$ *. Reactions are shown in Figure 5.3.7b. Considering the free-body diagrams in Figure 5.3.7c,d, we obtain the axial-force, shear-force, and bending-moment diagrams in Figure 5.3.8.

Because the bending moment is zero in each column, we reject Figure 5.3.6c,d with bent columns. We reject Figure 5.3.6e, in which the beam is concave down at its ends. Figure 5.3.6b is the plausible deflected shape because the beam bends concave up, consistently with the moment diagram, and the columns remain straight because there is no moment acting on them.

Let us determine the magnitude of the deflections and rotations in reference to Figure 5.3.9. For the loads applied, the bending moment in the beam is same as that

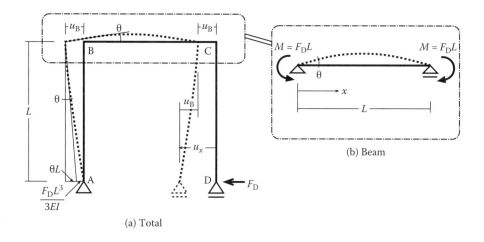

(a) Total

(b) Beam

FIGURE 5.3.5 Deflected shape.

---

* Note that we could have inferred this result directly from symmetry.

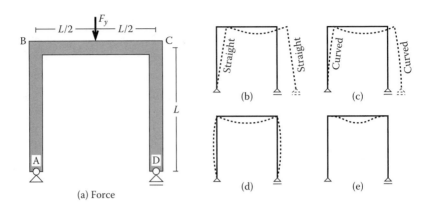

FIGURE 5.3.6 Portal frame with vertical force on beam.

of a simply supported beam (Figure 5.3.9b). Expressions for the deflection and the end slope of a simply supported prismatic beam with a concentrated load at midspan were given in Section 3.1, Chapter 3:

$$v = \frac{F_y L^3}{48EI} \quad (3.1.7)$$

$$\theta = \frac{F_y L^2}{16EI} \quad (3.1.4)$$

Because the columns are continuous with the beams at the joints, the slopes of the columns are also θ as shown in Figure 5.3.9a. With the pin support at A not being free to move horizontally, the horizontal displacement of the roller support D is

$$u_x = 2\theta L = \frac{F_y L^3}{8EI} \quad (5.3.4)$$

Check the result using GOYA-P.

To have a lateral movement of zero at the support, the inward movement caused by the lateral force F_D (Equation 5.3.3) should be offset by the outward movement

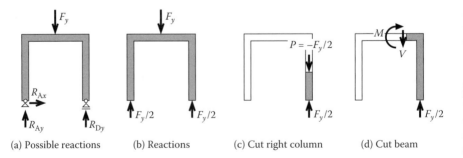

FIGURE 5.3.7 Reactions.

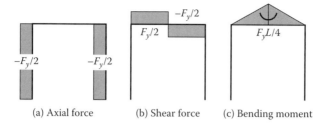

| (a) Axial force | (b) Shear force | (c) Bending moment |

FIGURE 5.3.8 Internal forces.

caused by the vertical force F_y (Equation 5.3.4). Therefore, we set the right-hand terms of the two equations equal to one another.

$$\frac{5F_D L^3}{3EI} = \frac{F_y L^3}{8EI} \tag{5.3.5}$$

from which we obtain the required proportion of F_D and F_y.

$$\frac{F_D}{F_y} = \frac{3}{40} \tag{5.3.6}$$

The axial-force, shear-force, and bending-moment diagrams are obtained by superposing 3/40 times the values in Figure 5.3.2 on those in Figure 5.3.8 as shown in Figure 5.3.10b–d.

The conditions depicted in Figure 5.3.10a are equivalent to those in Figure 5.3.11a where a portal frame is supported by two pins. The frame deforms (Figure 5.3.11b) in accordance with the bending-moment diagram in Figure 5.3.10d.

Assume that a horizontal load F_x is applied on the portal frame (Figure 5.3.12a). Select the correct deflected shape among those in Figure 5.3.12b–e.

The first step is to determine the reactions shown in Figure 5.3.13a. The given support conditions will allow horizontal and vertical reactions at A and a vertical

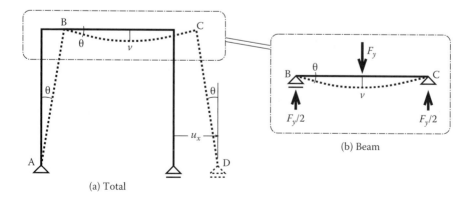

(a) Total

(b) Beam

FIGURE 5.3.9 Deflected shape of portal frame.

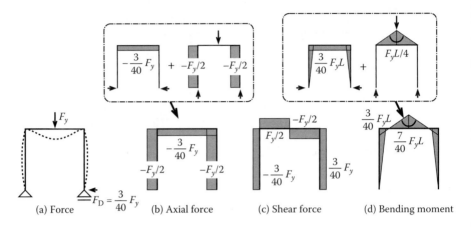

FIGURE 5.3.10 Reduce the horizontal displacement of the roller to zero.

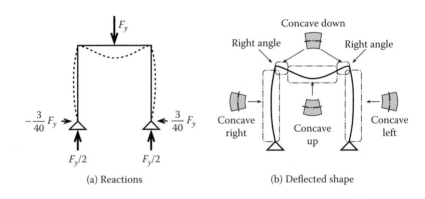

FIGURE 5.3.11 Portal frame with pin supports.

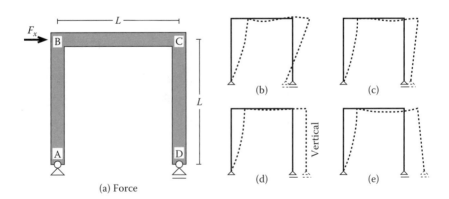

FIGURE 5.3.12 Portal frame with horizontal force.

Frames

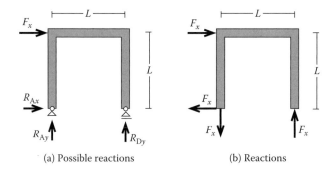

(a) Possible reactions (b) Reactions

FIGURE 5.3.13 Reactions.

reaction at D. We note that the height of the frame is equal to its span. The equilibrium of horizontal forces yields $R_{Ax} = -F_x$. The equilibrium of vertical forces and moments requires $R_{Ay} = -F_x$ and $R_{Ay} = F_x$. We conclude that the reactions are as shown in Figure 5.3.13b.

Considering the free-body diagrams (Figure 5.3.14a–c), we obtain the axial-force, shear-force, and bending-moment diagrams in Figure 5.3.15. We look at Figure 5.3.15c carefully. The moment diagram indicates that the left column and the beam should have compressive strain on their outside faces. They should bend so that they are concave out. The right column, not subjected to moment, should remain straight. We decide that the correct deflected shape is the one shown in Figure 5.3.12e. (Check using GOYA-P.)

Let us compute the deformations in reference to Figure 5.3.16a. Because the two ends of the beam are connected to the columns, vertical displacements at the two ends of the beam are considered to be negligible. Therefore, the beam bends as shown in Figure 5.3.16b. Recall the following equation in Chapter 2:

$$\frac{d^2v}{dx^2} = \frac{M}{EI} \tag{2.8.9}$$

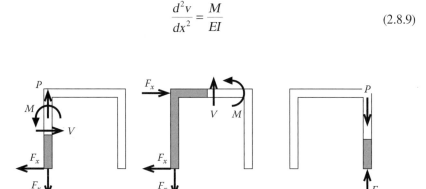

(a) Cut left column (b) Cut beam (c) Cut right column

FIGURE 5.3.14 Free-body diagrams.

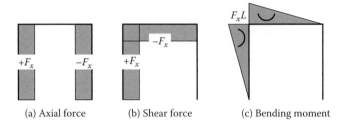

(a) Axial force (b) Shear force (c) Bending moment

FIGURE 5.3.15 Internal forces.

Substituting the bending moment shown in Figure 5.3.16b, $M = F_x \cdot (L - x)$, and integrating, we obtain

$$\frac{dv}{dx} = \frac{F_x}{EI}\left(xL - \frac{x^2}{2}\right) + C_1 \tag{5.3.7}$$

and

$$v = \frac{F_x}{EI}\left(\frac{x^2 L}{2} - \frac{x^3}{6}\right) + C_1 x + C_2 \tag{5.3.8}$$

The boundary condition that $v = 0$ at $x = 0$ and $x = L$ leads to

$$C_1 = -\frac{F_x L^2}{3EI} \quad \text{and} \quad C_2 = 0 \tag{5.3.9}$$

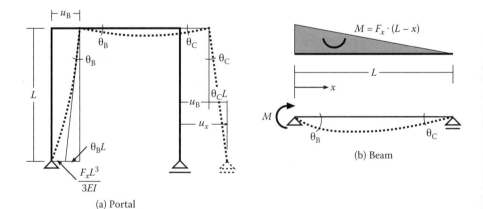

(a) Portal

(b) Beam

FIGURE 5.3.16 Deflected shape.

Frames

Substituting this into Equation 5.3.7, we obtain the slopes at beam ends:

$$\theta_B = -\left.\frac{dv}{dx}\right|_{x=0} = \frac{F_x L^2}{3EI} \quad \text{and} \quad \theta_C = \left.\frac{dv}{dx}\right|_{x=L} = \frac{F_x L^2}{6EI} \quad (5.3.10)$$

The horizontal displacement of the beam (u_B in Figure 5.3.16a) is obtained by adding the deformation of the left column and the contribution of the rotation:

$$u_B = \frac{F_x L^3}{3EI} + \theta_B L = \frac{2F_x L^3}{3EI} \quad (5.3.11)$$

The displacement of the roller (u_x in Figure 5.3.16a) is obtained by adding the displacement of the beam and the contribution of the rotation of the right column:

$$u_x = u_B + \theta_C L = \frac{5F_x L^3}{6EI} \quad (5.3.12)$$

Figure 5.3.17a shows the deformation of the portal frame subjected to a horizontal force F_x (Figure 5.3.13). Figure 5.3.17b shows the deformation of the portal subjected to a horizontal force $F_x/2$ on the roller support (see Equations 5.3.2 and 5.3.3). Adding the deflections in the two figures, we obtain Figure 5.3.17c, which refers to a portal frame supported by two pins and subjected to a horizontal force at the top. Check this result using GOYA-P. Superposing one half of Figure 5.3.2 and Figure 5.3.15, we obtain the axial-force, shear-force, and bending-moment diagrams in Figs. 5.3.18a–c. The bending-moment diagram indicates that the portal frame deforms as shown in Figure 5.3.18d.

EXAMPLE 5.3.1

Construct the bending-moment diagram for the portal frame shown in Figure 5.3.19.

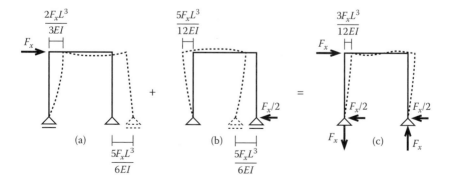

FIGURE 5.3.17 Reduce the horizontal displacement of the roller to zero.

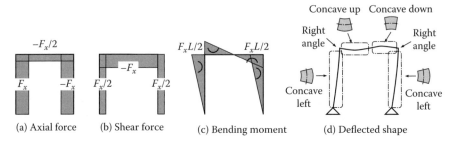

FIGURE 5.3.18 Portal frame with pin supports.

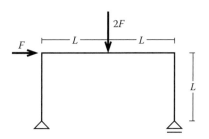

FIGURE 5.3.19 Portal frame.

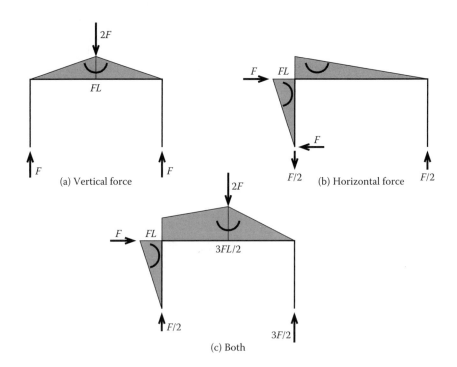

FIGURE 5.3.20 Bending moment diagrams.

Frames

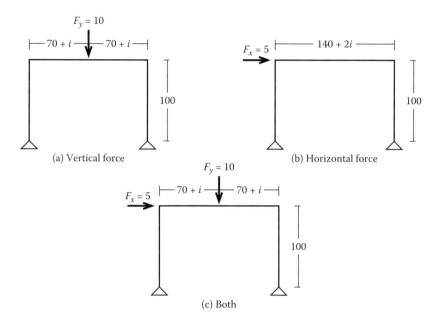

FIGURE 5.3.21 Portal frame with vertical (a), horizontal (b), and combined (c) loads.

Solution

The reactions and the bending moment caused by the vertical force are shown in Figure 5.3.20a. Those caused by the horizontal force are shown in Figure 5.3.20b. Superposing these figures, we obtain Figure 5.3.20c. (Determine the deflected shape using GOYA-P.)

EXERCISE

Select any one-digit number i. Construct the bending moment diagram for the portal frame shown in Figure 5.3.21 and sketch the deflected shape.

Hint: Reexamine the process leading to Equation 5.3.5 very carefully and project it to the case where the length of the beam is different from that of the columns.

5.4 STATICALLY INDETERMINATE FRAME

In Section 1.5, Chapter 1, we distinguished between statically determinate and statically indeterminate trusses: the axial forces in the statically determinate trusses can be calculated considering the equilibrium of forces only, but the calculation of the axial forces in the statically indeterminate trusses requires, in addition to statics, consideration of deformation compatibility. In this section, we shall study a statically indeterminate frame, for which the calculation of the internal forces (axial forces, shear forces, and bending moment) requires consideration of both force equilibrium and deformation compatibility. Many frames are statically indeterminate. We need to learn to determine their response to load. Compared with the calculation process for a determinate structure, that for an indeterminate structure is longer yet straightforward.

Recall Figure 1.5.9, Chapter 1, where the axial forces in statically indeterminate trusses changed when we changed the axial stiffness EA for the section. Similarly, the internal forces in an indeterminate frame changes if we change the bending stiffness EI (Young's modulus multiplied by the cross-sectional area) for the section. In Figure 5.3.10d, we obtained a bending-moment diagram for a portal frame with all of its members having the same stiffness. If we assume that the stiffness of column CD is extremely large as shown in Figure 5.4.1, which bending-moment diagram included in Figure 5.4.2a–d is the correct one?

We note that the moment diagram in Figure 5.4.2a is identical with the one obtained for a frame with all of its members having the same stiffness and, in Figure 5.4.2b,c, the columns resist larger moments than those in Figure 5.4.2a.

The portal frame in Figure 5.4.1 is an indeterminate frame. We cannot determine the internal forces using the conditions of equilibrium alone. So we decide to go through a simple three-step procedure.

Step 1: We release the horizontal-force restraint on support D (Figure 5.4.3a). Now the portal frame, with the degrees of freedom increased, is determinate. We can determine the reactions and the internal bending moment on the basis of equilibrium conditions alone (Figure 5.4.3b).

Given the distributions of bending moment, we can determine the horizontal displacement, u_{FD}, at support D (Figure 5.4.3c).

Step 2: Again referring to the frame with the released horizontal restraint, we apply a horizontal force of a given magnitude, R, at support D (Figure 5.4.4a). We get the bending moment diagram shown in Figure 5.4.4b and determine the horizontal displacement u_{RD} (Figure 5.4.4c).

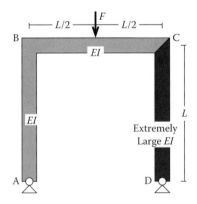

FIGURE 5.4.1 A portal frame with one flexible and very stiff column.

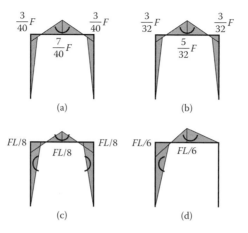

FIGURE 5.4.2 Options.

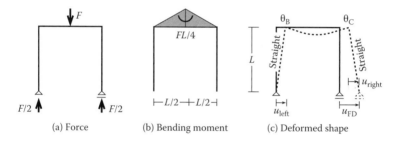

FIGURE 5.4.3 Vertical force on portal frame.

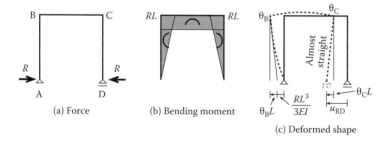

FIGURE 5.4.4 Horizontal force on portal frame.

Step 3: The condition we must satisfy is that the horizontal displacement at support D must be zero under the influence of the vertical load F and the horizontal load R. We determine the horizontal reaction at D (which must be equal to the horizontal reaction at A) from

$$u_{FD} = u_{RD} \tag{5.4.1}$$

We shall go through the previously described process in detail in the following paragraphs.

Step 1: If the horizontal restraint is released at support D, the portal frame has no horizontal reaction (Figure 5.4.3a). As we have observed earlier (Section 5.3), the beam responds as a simply supported beam (Figure 5.4.3b,c). Given that the slope is zero at midspan (deduced from symmetry), the rotation at each end is

$$\theta_B = \theta_C = \frac{FL^2}{16EI} \tag{5.4.2}$$

The columns are not subjected to bending moment. They do not bend. Therefore, the contributions of the column rotations to the horizontal deflections are $u_{\text{left}} = \theta_B L$ and $u_{\text{right}} = \theta_C L$ (because the column height is equal to the beam span L). The total displacement is

$$u_{\text{FD}} = u_{\text{left}} + u_{\text{right}} = \theta_B L + \theta_C L = \frac{FL^3}{8EI} \tag{5.4.3}$$

Step 2: We apply a horizontal force R at support D (Figure 5.4.4a). The resulting bending-moment distribution is shown in Figure 5.4.4b. We note that the bending deformation of the column on the right (with an extremely large bending stiffness EI) is negligible. The slopes at the two ends of the beam are the same:

$$\theta_B = \theta_C = \frac{RL^2}{2EI} \tag{5.4.4}$$

Therefore, the total horizontal displacement (Figure 5.4.4c) is the sum of the contribution of the column on the left

$$u_{\text{left}} = \theta_B L + \frac{RL^3}{3EI} = \frac{5RL^3}{6EI} \tag{5.4.5}$$

and that of the very stiff column on the right

$$u_{\text{right}} = \theta_C L = \frac{RL^3}{2EI} \tag{5.4.6}$$

resulting in

$$u_{\text{RD}} = u_{\text{left}} + u_{\text{right}} = \frac{4RL^3}{3EI} \tag{5.4.7}$$

Step 3: To get the proper horizontal reaction at D, we equate the displacements ($u_{\text{FD}} = u_{\text{RD}}$)

$$\frac{FL^3}{8EI} = \frac{4RL^3}{3EI} \tag{5.4.8}$$

leading to

$$R = \frac{3}{32}F \qquad (5.4.9)$$

Now we have the horizontal reaction for the indeterminate frame in terms of the force F. Because we know the reactions (or external forces), we determine the bending moment at the top of the column (which is equal to the moment at the end of the beam) using statics

$$M = R \times L = \frac{3}{32}FL \qquad (5.4.10)$$

We conclude that solution 5.4.2b is the correct diagram.

The process can be repeated in GOYA-P as follows:

1. Make the vertical force zero and apply a horizontal force of –3N at the roller support (support B).
2. Click the "Setting" button and increase E for the right column to at least 100 times the default value. Note that the horizontal displacement of the roller has now been reduced to 4/5 of that in the previous stage.
3. Apply a vertical force of 32 N at midspan of the beam. You will find that Equation 5.4.9 is satisfied and that the horizontal displacement of the roller support is zero.

What would happen if you increase EI for both columns equally to an extremely large value? In this case, the displacement of the roller caused by the horizontal force R (Figure 5.4.4a) is

$$u_{RD} = \theta_B L + \theta_C L = \frac{RL^3}{EI} \qquad (5.4.11)$$

To get the proper horizontal reaction at D, we equate this displacement to that caused by the vertical force (Equation 5.4.3).

$$\frac{RL^3}{EI} = \frac{FL^3}{8EI} \qquad (5.4.12)$$

which leads to

$$R = \frac{1}{8}F \qquad (5.4.13)$$

Therefore, the bending moment is as shown in Figure 5.4.2c. The deflected shape is shown in Figure 5.4.5. Note that the columns do not deflect. The deflected shape and bending moment of the beam are, therefore, the same as those of a beam with both ends fixed as shown in Figure 5.4.6.

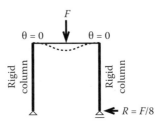

FIGURE 5.4.5 Two rigid columns.

EXAMPLE 5.4.1

Construct the bending-moment diagram of the structure shown in Figure 5.4.7a and calculate the horizontal displacement of the beam, u_x.

Solution

Because the beam is rigid and does not deform, the top of each column cannot rotate ($\theta = 0$) (Figure 5.4.7b). Therefore, the columns are deformed as if they are cantilever beams of length L with their free ends at the supports. Because the stiffness (EI) and the lateral displacement (u_x) of the two columns are the same, the shear forces in the two columns are the same. We conclude that the shear force in each column is $F/2$ and the bending moment diagram is as shown in Figure 5.4.7c. Note that the deflection of a cantilever beam with a load of $F/2$ is

$$u_x = \frac{(F/2)L^3}{3EI} = \frac{FL^3}{6EI}$$

This is the lateral displacement of the beam shown in Figure 5.4.7b. In GOYA-P, modify Young's modulus of the columns to 178 N/mm² so that $EI = 10^6$ N.mm², and make Young's modulus of the beam very large (at least 100 times the default value). Make $F = 6$ N. The lateral displacement of the beam indicated on the screen should be $u_x = 6 \times 100^3 / (6 \times 10^6) = 1$ mm.

EXAMPLE 5.4.2

Draw the bending-moment diagram of the structure shown in Figure 5.4.8a and calculate the horizontal displacement of the beam, u_x.

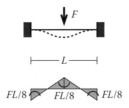

FIGURE 5.4.6 Beam with fixed ends.

Frames

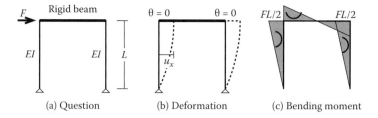

FIGURE 5.4.7 A symmetric frame with a rigid beam.

Solution

Because the beam is rigid, the columns may be treated as cantilever beams with their fixed ends at the beam level. Let R_1 and R_2 denote the reactions of the supports as shown in Figure 5.4.8b and note that the deformation (u_x) of each column is the same

$$u_x = \frac{R_1 L^3}{3EI} = \frac{R_2 L^3}{3 \times (2EI)} \tag{5.4.14}$$

or $R_2 = 2R_1$. Noting that $F = R_1 + R_2$, we obtain $R_1 = F/3$ and $R_2 = 2F/3$. The shear force in the right column is twice that in the left column because the bending stiffness of the right column is twice that of the left column. If we substitute $R_1 = F/3$ into Equation 5.4.14, we get

$$u_x = \frac{(F/3)L^3}{3EI} = \frac{FL^3}{9EI} \tag{5.4.15}$$

You should check this result using GOYA-P.

EXAMPLE 5.4.3

Construct the bending-moment diagram of the structure with distributed load shown in Figure 5.4.9.

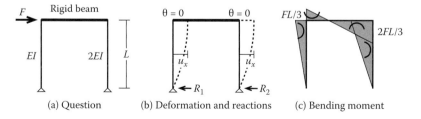

FIGURE 5.4.8 An unsymmetrical frame with a rigid beam.

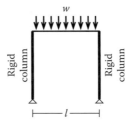

FIGURE 5.4.9 A frame with distributed load.

Solution

We can solve this example replacing one of the pin supports by a roller support as we did before. However, we note that the columns do not deform. That observation leads to a faster solution. Because the ends of the beam do not rotate, the beam may be considered a beam with fixed ends (Figure 5.4.10a).

Note that the vertical reaction at each support is $wL/2$ because of symmetry. Integrating $dV/dx = -w$, we obtain Figure 5.4.10b or

$$V = \frac{wL}{2} - wx \tag{5.4.16}$$

The bending moment is obtained integrating $dM/dx = V$:

$$M = M_0 + \frac{wL}{2}x - \frac{w}{2}x^2 \tag{5.4.17}$$

where M_0 denotes the bending moment at $x = 0$ (the left end). We substitute this equation into $d^2v/dx^2 = M/EI$ and integrate it, noting that the inclination at $x = 0$ (the left end)

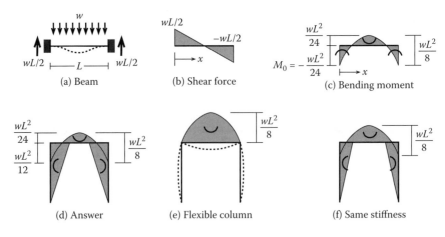

FIGURE 5.4.10 Solution.

is zero. We obtain

$$\frac{dv}{dx} = \frac{1}{EI}\left(M_0 x + \frac{wL}{4}x^2 - \frac{w}{6}x^3\right) \quad (5.4.18)$$

Because the deflected shape is symmetrical about midspan, the slope at $x = L/2$ (midspan) must be also zero. Therefore,

$$M_0 = -\frac{wL^2}{12} \quad (5.4.19)$$

Figure 5.4.10c shows the bending-moment diagram.

If the columns are much more flexible than the beam (contrary to Figure 5.4.9), the bending moments at the ends of the beam M_0 approach zero and we obtain the bending-moment diagram as shown in Figure 5.4.10e that is equivalent to that of a simple beam. If the stiffness of the beam is similar to those of the columns, the bending-moment diagram is between those shown in Figure 5.4.10d and e, as shown in Figure 5.4.10f.

In Figure 5.4.10d–f, you should note that the difference between the moment at the ends and that at midspan is always $wL^2/8$. This agrees with the bending moment at midspan of a simple beam under a uniform load of w (see Section 3.2, Chapter 3). Such an agreement is also observed for the case of a concentrated load (Figure 5.4.2a-c). The bending moment in an equivalent simple-beam is called "static moment." Figure 5.4.11a shows a frame subjected to a uniform load of w and a horizontal load of F_H. If you call the positive moment at midspan M_1 and the two negative moments at the ends of the beam M_0 and M_2, you will find

$$M_1 - \frac{M_0 + M_2}{2} = \frac{wL^2}{8} \quad \text{or} \quad M_1 + \frac{|M_0 + M_2|}{2} = \frac{wL^2}{8}$$

because the moment distribution in Figure 5.4.11a can be decomposed into those in Figure 5.4.11b,c. For a beam with a concentrated load at the middle (Figure 5.4.12),

$$M_1 - \frac{M_0 + M_2}{2} = \frac{FL}{4} \quad \text{or} \quad M_1 + \frac{|M_0 + M_2|}{2} = \frac{FL}{4}$$

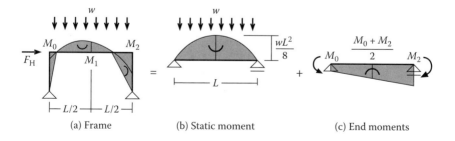

(a) Frame (b) Static moment (c) End moments

FIGURE 5.4.11 Frame subjected to distributed load and horizontal load.

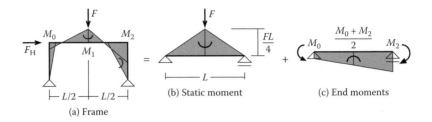

FIGURE 5.4.12 Frame subjected to vertical and horizontal loads.

The left-hand term, for any loading distribution, is equal to the static moment or the moment at midspan of a simply supported beam for the same loading distribution. This result is useful to remember because it can be used to check the reliability of solutions for statically indeterminate beams.

EXERCISE I

Choose any one-digit number i to determine the length of the beam in Intext Figure 5.4.13. Fill out the table using GOYA-P while keeping Young's modulus of the other members as the default value ($E = 100$ N/mm²). Also, sketch the deflected shape of the frame for each case.

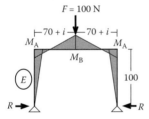

E of Left Column	R	M_A	M_B
500 N/mm²			
100 N/mm²			
20 N/mm²			

FIGURE 5.4.13 Unsymmetrical frame subjected to vertical load.

EXERCISE 2

Choose any one-digit number i to determine the length of the beam in Intext Figure 5.4.14. Fill out the table using GOYA-P but keep Young's modulus of the other members equal to the default value ($E = 100$ N/mm²). Sketch the deflected shape of the frame for each case.

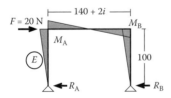

E of Left Column	R_A	R_B	M_A	M_B
500 N/mm²				
100 N/mm²				
20 N/mm²				

FIGURE 5.4.14 Unsymmetrical frame subjected to horizontal load.

Frames

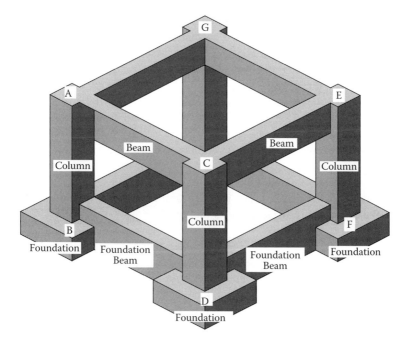

FIGURE 5.5.1 Structural concrete building.

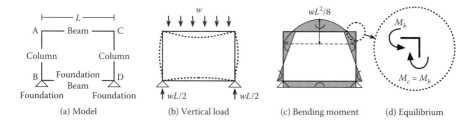

FIGURE 5.5.2 One-story one-span frame under vertical load.

5.5 MULTISTORY FRAME

GOALS OF STRUCTURAL DESIGN

In the First century B.C., a Roman builder, Vitruvius, wrote that a building should be strong, that it should function properly, and that it should please. We translate those maxims into the engineer's current work environment as follows:

1. Safety: A building structure must have the strength to resist permanent and transient loads* with appropriate factors of safety. It must not endanger the lives of its occupants.

---

* The main source of permanent loads is gravity. Transient loads are caused by such effects as traffic, wind and earthquakes.

2. **Serviceability:** The deformations of a building structure, in the short or long term, must be controlled so that its functionality is not impaired. The engineer needs to consider not only stresses and strains but also durability.
3. **Beauty:** A building structure should be planned and proportioned to please the eye. Except for transportation and industrial structures, this task is seldom within the purview of the structural designer. Nevertheless, the structural designer must always consider this important aspect of the design challenge. Structures, as well as being right, need to look right.

Structural design is essentially an art. In the practice of engineering, one rarely obtains a unique answer from a crisp design procedure. Even in the simplest of cases, structural design requires trial and error and involves a two-step process:

Step 1: Determine the framing and dimensions of the structure. This step requires not only theoretical knowledge but also the ability to observe and weigh relevant evidence. An engineer should never accept an answer from a calculation that the engineer could not have guessed at to within +/− 20%. Before the design calculations are initiated, the engineer should be able to make good estimates of the sizes of the structural elements. The student may well ask, "How does one do that if one has never designed a structure?" The student can compensate for his or her lack of experience by looking at structures critically and developing a sensitivity to ratios such as beam span to depth and floor to column-section area.

Step 2: Check if the selected framing and dimensions satisfy the requirement of serviceability and safety. The most effective way to satisfy this requirement would be to build the structure and observe its behavior under service or test loads. For most civil-engineering structures, such an approach would be prohibitively expensive. So, we "test" the structure by calculation, and if the calculation requires much labor, we use computers. Numerical models of the structure, based on methods we have been discussing and have implemented using powerful computers, give us "seven-league boots" to help decide whether the sizes we have selected are satisfactory. However, in doing that, we must be very careful to make certain that the conclusions fit into our sense of proportion. Every time we obtain an answer through a sophisticated and complicated procedure, it is our professional responsibility to try to obtain the same answer, if approximately, through simple thinking. The simplest way to think is to know the proportions of the answer from previous observation. That is why experience is important in engineering.

ILLUSTRATIVE DESIGN EXAMPLES

The designing of a structure requires extensive work and great care for detail. We cannot undertake that in the classroom. The following examples will illustrate certain basic points of the dimensioning process for a frame. In that respect, they are not design examples but illustrations. Our goal is to understand how the internal forces, which will subsequently be used in determining the properties of the elements, are determined.

Most modern buildings in developed countries are made of concrete, steel, or timber. Figure 5.5.1 shows a very simple example of the framing of a structural reinforced concrete building. The gravity load on the floor (not shown) is transmitted to beams. The beams transmit the load to the columns. The columns rest on the foundations, and the foundations rest on the ground. The sequence of the gravity load going from the slab to the girders to the columns to the foundations is called the "load path." It is important in design to have a good grasp of the load path.

We can model a portion of this building (frame ABCD in Figure 5.5.2a). The foundations are modeled as pin supports because the ground under the foundations is usually soft and its resistance to rotation is much smaller than that of the foundation beams. The gravity load on the slab is transferred to the beam almost uniformly as shown in Figure 5.5.2b. This load induces the bending moments shown in Figure 5.5.2c. The moment diagrams in Figure 5.5.2c are similar to those in Figure 5.4.10f, except that the bending moment at the bottom of each column is not zero because of the resistance of the foundation beam. Note that bending moments are in equilibrium at each joint (or *node*) as shown in Figure 5.5.2d. The other frames (such as CDEF) can be modeled similarly. As a result, each column is bent in two directions.

Figure 5.5.3a shows a symmetrical frame with two spans subjected to uniform vertical load. Because of the symmetry, the column in the middle does not bend. It is not subjected to moment because the end moments of the beams balance one another (Figure 5.5.3b). The bending moment in the foundation beams is negative near the middle because the foundations are assumed to be pin supports that do not move. (In reality, the foundations move as the ground deforms, but the deformation of the ground is usually very small.)

Figure 5.5.4a shows a two-story frame with a span subjected to uniform vertical load. Figure 5.5.4b shows the bending-moment diagram. Note that at each end node of the beam CD (Figure 5.5.4c), equilibrium of moment requires $M_b = M_{c1} + M_{c2}$. The bending moment in the beam at the node is about twice that of each column.

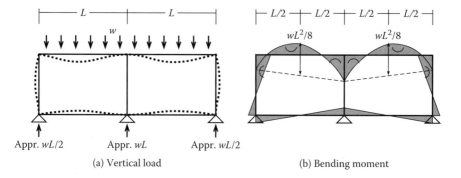

FIGURE 5.5.3 One-story two-bay frame under vertical load.

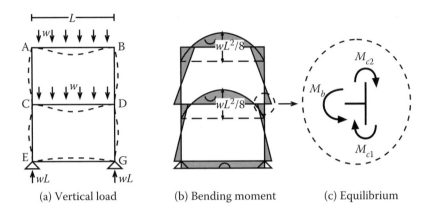

(a) Vertical load (b) Bending moment (c) Equilibrium

FIGURE 5.5.4 Two-story one-bay frame under vertical load.

Figure 5.5.5a shows a one-story one-bay frame subjected to a horizontal load. If the stiffnesses of the columns are the same, the shear forces in the columns are also the same ($V = F/2$) as discussed in Section 5.4. The vertical reactions, $(H/L)F$, are determined by equilibrium conditions and are equal to those of a portal frame without a foundation beam (Figure 5.4.7). Because foundation beam CD is designed to be stiffer than beam AB, the bending moment at the bottom of the column is larger than the moment at the top of the column (Figure 5.5.5b). Note again that bending moments are in equilibrium at each node (Figure 5.5.5c).

Figure 5.5.6a shows a one-story two-bay frame subjected to a horizontal load. The bending-moment diagram (Figure 5.5.6b) is similar to that of Figure 5.5.5b, except that the column in the center resists a larger bending moment than the other columns because the column in the center is connected to two beams as shown in Figure 5.5.6c. Equilibrium requires $M_c = M_{b1} + M_{b2}$.

Figure 5.5.7a shows a two-story one-bay frame subjected to horizontal load. The bending moment diagram is shown in Figure 5.5.7b. As required by equilibrium of moments shown in Figure 5.5.7c ($M_b = M_{c1} + M_{c2}$), the bending moment in beam CD is larger than that in roof beam AB. If the flexural stiffness of the left column is

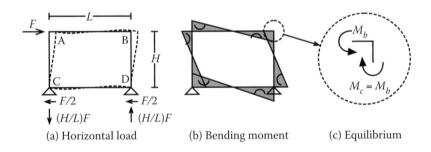

(a) Horizontal load (b) Bending moment (c) Equilibrium

FIGURE 5.5.5 One-story one-bay frame under horizontal load.

Frames

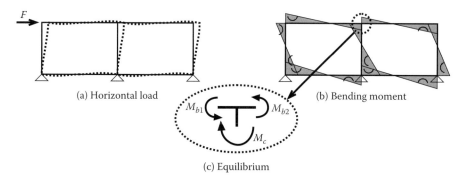

FIGURE 5.5.6 One-story two-bay frame under horizontal load.

the same as that of the right column, the shear forces of the columns in the second story are $V = F_2/2$ as shown in Figure 5.5.7d. On the other hand, the shear forces of the columns in the first story are $V = (F_1 + F_2)/2$ (Figure 5.5.7e).

EXAMPLE 5.5.1

A two-story frame is subjected to horizontal loads, and the columns resist bending moments as shown in Figure 5.5.8.

1. Compute the shear forces in the beams.
2. Compute the axial forces in the columns.
3. Compute the applied horizontal forces and the reactions.

Solution

1. If we consider equilibrium at each node, we obtain the bending-moment diagram for each beam as shown in Figure 5.5.9. Because the slope of the bending-moment diagram is equal to the shear force, we conclude that the shear force in

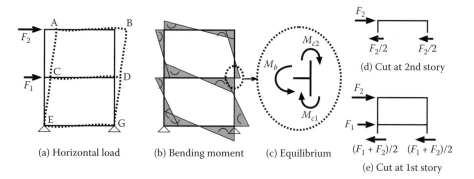

FIGURE 5.5.7 Two-story one-bay frame under horizontal load.

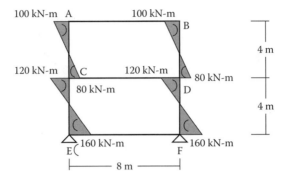

FIGURE 5.5.8 Bending moment in columns.

each beam is

$$\text{Beam AB} \quad V = \frac{2 \times 100}{8} = 25 \text{ kN}$$

$$\text{Beam CD} \quad V = \frac{2 \times 200}{8} = 50 \text{ kN}$$

$$\text{Beam EF} \quad V = \frac{2 \times 160}{8} = 40 \text{ kN}$$

2. If we cut the frame at the second story and at midspan of roof beam AB as shown in Figure 5.5.10a, we can conclude that the axial force in column AC is $P = 25$ kN in tension. If we cut the frame at the first story as shown in Figure 5.5.10b, we can conclude that the axial force in column CE is $P = 25 + 50 = 75$ kN in tension. The axial forces in columns BD and DF have the same magnitude as columns AC and CE, respectively, but they are in compression.

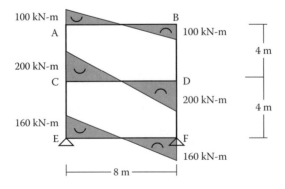

FIGURE 5.5.9 Bending moment in beams.

Frames

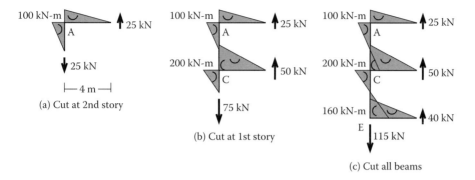

FIGURE 5.5.10 Cut the frame.

3. If we cut all the beams as shown in Figure 5.5.10c, we can determine that the vertical reaction at support E is

$$R = 25 + 50 + 40 = 115 \text{ kN}$$

The vertical reaction at support F has the same magnitude but is in the opposite direction (Figure 5.5.11). We can compute the shear forces in the columns using the slope of the bending-moment diagram shown in Figure 5.5.8:

$$\text{Columns AC and BD} \quad V = \frac{100 + 80}{4} = 45 \text{ kN}$$

$$\text{Columns CE and DF} \quad V = \frac{120 + 160}{4} = 70 \text{ kN}$$

Recalling Figs. 5.5.7d,e, we conclude that the horizontal force on the roof is $45 \times 2 = 90$ kN. On the second floor, it is $70 \times 2 - 90 = 50$ kN (Figure 5.5.11). The magnitudes of the horizontal reactions are same as the shear forces in the columns in the first story.

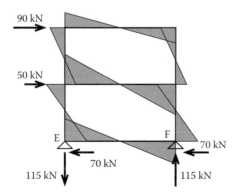

FIGURE 5.5.11 Forces and reactions.

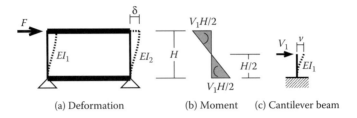

(a) Deformation (b) Moment (c) Cantilever beam

FIGURE 5.5.12 One-story frame with rigid beams.

Often, we use computers to calculate deformations of frames subjected to horizontal forces. If we assume that the flexural deformation of beams and the axial deformation of columns are negligible, we can approximate the deformation of frames without computers. Consider the frame shown in Figure 5.5.12a. Because the beams are assumed to be rigid, the bending moment at midspan of each column is zero as shown in Figure 5.5.12b. Figure 5.5.12c shows an equivalent cantilever beam with the flexural rigidity of the left column (EI_1) subjected to the shear force in the left column (V_1). Recalling Section 2.8 in Chapter 2, we can calculate the deflection of the beam as

$$v = \frac{(H/2)^3}{3EI_1} V_1 \tag{5.5.1}$$

Note that the deflection of the frame is twice that of the equivalent beam ($\delta = 2v$), which leads to

$$V_1 = \frac{12EI_1}{H^3} \delta \tag{5.5.2}$$

We can develop a similar equation for the right column.

$$V_2 = \frac{12EI_2}{H^3} \delta \tag{5.5.3}$$

Because the sum of the shear forces in the columns is equal to the external horizontal force ($F = V_1 + V_2$), we get

$$F = \frac{12(EI_1 + EI_2)}{H^3} \delta \quad \text{and} \quad \delta = \frac{H^3}{12(EI_1 + EI_2)} F \tag{5.5.4}$$

Note that the deformation δ determined by Equation 5.5.4 is smaller than the actual deformation.

Figure 5.5.13 shows a two-story frame subjected to horizontal loads. The deformation of each story (δ_1 and δ_2) is called "story drift." Using the procedure similar to that for one-story frame, we get

$$\delta_2 = \frac{H^3}{12(EI_{21} + EI_{22})} F_2 \quad \text{and} \quad \delta_1 = \frac{H^3}{12(EI_{11} + EI_{12})} (F_1 + F_2) \tag{5.5.5}$$

Note that the horizontal displacement of the roof is equal to the sum of the story drifts ($\delta_1 + \delta_2$).

Frames 279

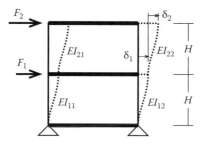

FIGURE 5.5.13 Two-story frame.

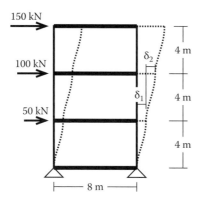

FIGURE 5.5.14 Three-story frame.

EXERCISE USING GOYA-M

Assume a three-story frame shown in Figure 5.5.14, where the cross sections of the columns in the second and third story measure 500 × 500 mm, and Young's modulus is 20 kN/mm². You may neglect the flexural deformation of beams and the axial deformation of columns.

1. Construct the bending-moment diagram of beams and columns.
2. Estimate the drift of the second story.
3. Determine the cross-sectional dimensions of the columns in the first story so that the drift of the first story is the same as that of the second story.
4. Check your results using GOYA-M.

5.6 THREE-HINGED FRAME

Suppose we construct a large portal frame in October (Figure 5.6.1a). In winter, the roof tends to shorten as shown in Figure 5.6.1b. As a result, the beam is bent down and subjected to a tensile axial force (Figure 5.6.1d,e). In summer, the roof is heated and tends to elongate as shown in Figure 5.6.1c. As a result, the beam is bent up and

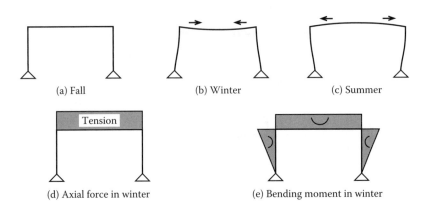

FIGURE 5.6.1 A portal frame.

subjected to a compressive axial force. If we build a hinge into the beam as shown in Figure 5.6.2a, such internal forces do not develop because the frame can deform freely as shown in Figs. 5.6.2b,c. Such a structure is called a *three-hinged frame* (one hinge in the beam and two at the supports).

The three-hinged frame in Figure 5.6.2 is statically determinate. Start GOYA-H and apply a vertical load at midspan. You will get the results shown in Figure 5.6.3. The shear-force and bending-moment diagrams are obtained considering equilibrium alone. Figure 5.6.4a shows the reactions at the supports. We assume that the span length is $2L$, the column height is L, and the frame is symmetrical. Because of symmetry, the magnitude of each vertical reaction (R_{Ay} and R_{Cy}) is $F/2$. Figure 5.6.4b shows the free-body diagram for frame segment AB. Moment equilibrium at support A leads to $R_{Bx} = F/2$. Moment equilibrium at hinge B leads to $R_{Ax} = F/2$. The shear-force and bending-moment diagrams are shown in Figure 5.6.4c,d.

What if we move the load to the right as shown in Figure 5.6.5a? Moment equilibrium at support A leads to $R_{Cy} = 3F/4$. Moment equilibrium at support C leads to $R_{Ay} = F/4$. Figure 5.6.5b shows the free-body diagram for frame segment AB. Force equilibrium in the vertical direction leads to $R_{By} = F/4$. Moment equilibrium at hinge B leads to $R_{Ax} = F/4$. Moment equilibrium at support A leads to $R_{Bx} = F/4$. The shear-force and bending-moment diagrams are shown in Figure 5.6.5c,d. Figure 5.6.6a shows the free-body diagram of frame segment BC. Moment equilibrium at hinge B leads to $R_{Cx} = F/4$. The shear-force and

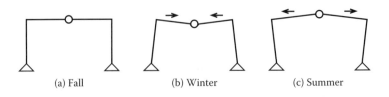

FIGURE 5.6.2 Three-hinged frame.

Frames

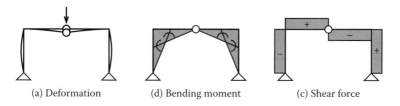

FIGURE 5.6.3 Vertical load at midspan.

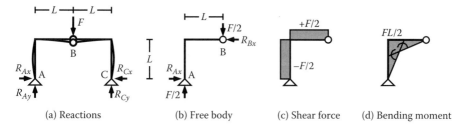

FIGURE 5.6.4 Equilibrium.

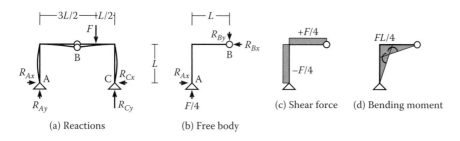

FIGURE 5.6.5 Move the load to the right.

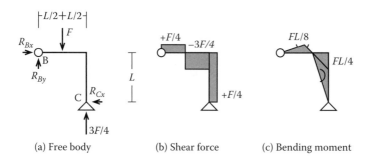

FIGURE 5.6.6 Equilibrium of frame segment BC.

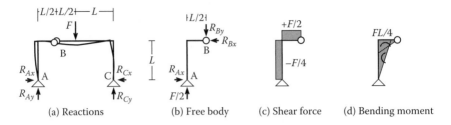

FIGURE 5.6.7 Move the hinge to the right.

bending-moment diagrams are in Figure 5.6.6b,c. As we move the location of the vertical load to the right, the bending moment in the beam decreases as does the deformation of the frame. If we move the location of the load so that it coincides with the column axis, both the bending moment and the beam deflection become zero.

Move the location of the load back to midspan and move the location of the hinge to the left as shown in Figure 5.6.7a. This move also changes the shear-force and bending-moment diagrams. Moment equilibrium of the structure in Figure 5.6.7a at support A or C leads to $R_{Ay} = R_{Cy} = F/2$ as they were before we moved hinge B. Figure 5.6.7b shows the free-body diagram for segment AB. Force equilibrium in the vertical direction leads to $R_{By} = F/2$. Moment equilibrium at hinge B leads to $R_{Ax} = F/4$. Moment equilibrium at support A leads to $R_{Bx} = F/4$. We obtain the shear-force and bending-moment diagrams in Figure 5.6.7c,d. Figure 5.6.8a shows the free-body diagram for frame segment BC. Moment equilibrium at hinge B leads to $R_{Cx} = F/4$. Thus, we get the diagrams in Figure 5.6.8b,c. If we move the location of the hinge to the left end, the bending moments in the columns diminish. The beam is bent in the same way as a cantilever beam. Because the flexural deformation in the columns is zero, the beam displaces to the left. We can also notice similar displacement in Figure 5.6.7a.

Using GOYA-H, apply a horizontal load F_x on the frame as shown in Figure 5.6.9a. Moment equilibrium at support A or C leads to $R_{Ay} = R_{Cy} = F_x/2$. Figure 5.6.9b shows the free-body diagram for frame segment AB. Force equilibrium in the vertical direction leads to $R_{By} = F_x/2$. Moment equilibrium at hinge B leads to $R_{Ax} = F_x/4$. Force equilibrium in the horizontal direction leads to $R_{Bx} = F_x - R_{Ax} = 3F_x/4$. These results help us obtain the diagrams in Figure 5.6.9c,d. Figure 5.6.10a shows the free-body

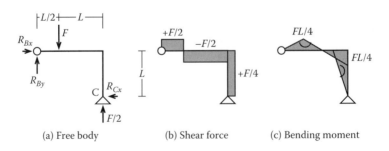

FIGURE 5.6.8 Equilibrium of frame segment BC.

Frames

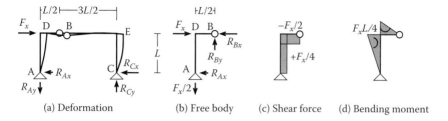

FIGURE 5.6.9 Horizontal load.

diagram for frame segment BC. Force equilibrium in the horizontal direction leads to $R_{Cx} = R_{Bx} = 3F_x/4$ to obtain the diagrams in Figure 5.6.10b,c. Note that the bending moment in column CE is larger than that in column AD. If we move the location of the hinge to the left end, the bending moments of column AD are reduced.

EXAMPLE 5.6.1

Figure 5.6.11 shows a three-hinged frame subjected to horizontal and vertical loads, F_x and F_y. Assume that $F_y = 30$ kN and the bending moment at node E is zero. Compute the required magnitude of F_x and construct the moment diagram.

Solution

Using the results obtained in Figures 5.7.7d and 5.6.8c, we determine the moment diagram caused by $F_y = 30$ kN (Figure 5.6.12a). Using the results from Figures 5.6.9d and 5.6.10c, we determine the moment diagram caused by F_x (Figure 5.6.12b). Because the bending moment at node E is zero, we have the condition $60 + 6F_x = 0$ or $F_x = -10$ kN. Figure 5.6.12c shows the moment diagram. Check the result using GOYA-H. You will find that the horizontal displacement of the beam is zero.

We can obtain the same solution without using the superposition technique. Figure 5.6.13a shows the possible reactions. Because the bending moment at node E was set to zero by choosing F_x, we conclude that the shear force in column EC and the reaction R_{Cx} are zero. Force equilibrium in the horizontal direction ($F_x = R_{Ax} + R_{Cx}$) leads to $R_{Ax} = F_x$. Moment equilibrium at support C leads to

$$8F_x + 16R_{Ay} - 8 \times 30 = 0 \quad \text{or} \quad R_{Ay} = \frac{30 - F_x}{2}$$

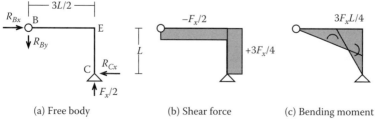

FIGURE 5.6.10 Equilibrium of frame segment BC.

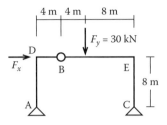

FIGURE 5.6.11 Three hunged frame subjected to horizontal and vertical loads.

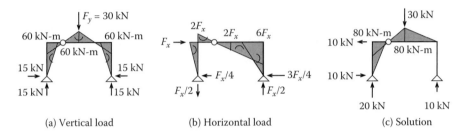

FIGURE 5.6.12 Solution using the previous results.

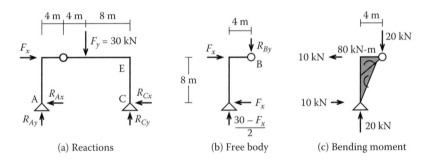

FIGURE 5.6.13 Another solution.

Figure 5.6.13b shows the free-body-diagram for frame segment AB. Moment equilibrium at hinge B leads to

$$8F_x + 4 \times \frac{30 - F_x}{2} = 0 \quad \text{or} \quad F_x = -10 \, \text{kN}$$

Force equilibrium in the vertical direction leads to

$$R_{By} = \frac{30 - F_x}{2} = 20 \, \text{kN}$$

Thus, we obtain the moment diagram shown in Figure 5.6.13c. The moment diagram for frame segment BC is obtained similarly.

Frames

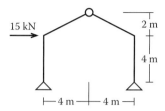

FIGURE 5.6.14 Gable frame with a hinge at its apex.

EXAMPLE 5.6.2

Construct the moment diagram for the frame shown in Figure 5.6.14.

Solution

Figure 5.6.15a shows possible reactions. Moment equilibrium at support A or E leads to $R_{Ay} = R_{Ey} = 7.5$ kN. Figure 5.6.15b shows the free-body diagram for frame segment CDE. Moment equilibrium at hinge C leads to $7.5 \times 4 - R_{Ex} \times 6 = 0$ or $R_{Ex} = 5$ kN. Thus, we obtain the moment diagram shown in Figure 5.6.15c. The moment diagram for frame segment ABC is obtained similarly. Figure 5.6.15d shows the moment diagram for the entire frame.

EXERCISE

Use a one-digit number i to define the vertical load F_y shown in Figure 5.6.16. Assume that the beams and columns can resist bending moment up to 500 kN-m. Compute the maximum (positive) and minimum (negative) horizontal loads that the frame can resist. Check the results using GOYA-H. Sketch the moment diagram and the deflected shape of the frame.

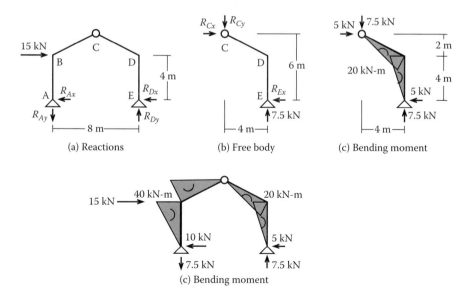

FIGURE 5.6.15 Solution.

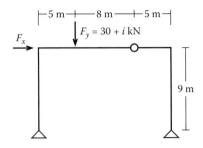

FIGURE 5.6.16 Three-hinged frame subjected to horizontal and vertical loads.

5.7 PROBLEMS

5.1 Figure 5.7.1 shows a bent subjected to three forces and an unknown couple, M. Determine the magnitude of the couple M to make the moment at the fixed-base A zero. Check to see if your answer is the same as one of the five options listed in Table 5.7.1. Positive sign refers to counterclockwise moment as shown in Figure 5.7.1.

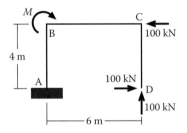

FIGURE 5.7.1 Bent.

TABLE 5.7.1

	M
1.	+ 1,000 kN-m
2.	+ 600 kN-m
3.	+ 400 kN-m
4.	− 600 kN-m
5.	− 1,000 kN-m

5.2 Figure 5.7.2 shows a bent subjected to horizontal and vertical forces, F_x and F_y. The bending moment at point A is to be zero. Select the correct ratio of F_x to F_y among the five options listed in Table 5.7.2 to make the moment at point A zero.

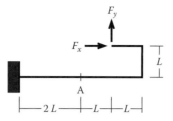

FIGURE 5.7.2 Bent.

TABLE 5.7.2

	$F_y : F_x$
1.	1:0
2.	1:1
3.	1:2
4.	3:3
5.	3:4

5.3 Figure 5.7.3 shows a portal subjected to horizontal and vertical forces, F. Select the correct bending-moment diagram among the options shown in Figure 5.7.4a–e.

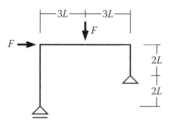

FIGURE 5.7.3 Portal.

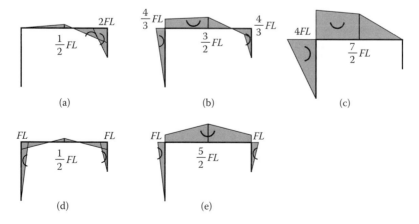

FIGURE 5.7.4 Options.

5.4 Figure 5.7.5 shows a portal frame supported on two hinges and subjected to a vertical load. Assume that the stiffness term, *EI*, for the beam and the columns is the same. Select the correct bending-moment diagram among the options shown in Figure 5.7.6a–e. (Hint: Construct a shear-force diagram that corresponds to each bending-moment diagram. Then, think!)

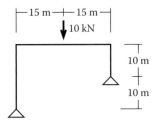

FIGURE 5.7.5 Portal.

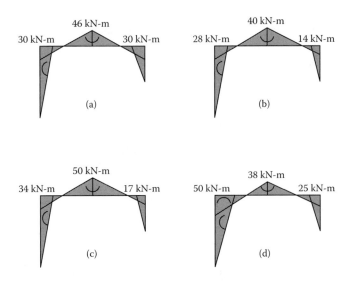

FIGURE 5.7.6 Options.

5.5 Figure 5.7.7 shows bending-moment diagrams for columns in a two-story frame subjected to lateral loads. Select the *incorrect* statement among the following options.
1. The shear force in beam AB is $V_{AB} = 35$ kN.
2. The shear force in beam CD is $V_{CD} = 62.5$ kN.
3. The axial force in column DG is $P_{DG} = 97.5$ kN.
4. The vertical reaction at support G is $R_{Gy} = 140$ kN.
5. The horizontal force on node D is $F_D = 160$ kN.

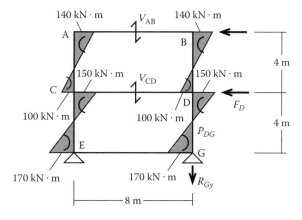

FIGURE 5.7.7 Two-story frame.

5.6. Figure 5.7.8 shows a two-story frame with horizontal loads. Assume that the flexural and axial deformation of the beams and the axial deformation of the columns are negligible. Assume also that the stiffness, EI, of the columns in the first story is twice that in the second story. Select the correct ratio of the drift in the first story, δ_1, to that in the second story, δ_2, from the options listed in Table 5.7.3.

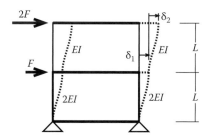

TABLE 5.7.3

	$\delta_1 : \delta_2$
1.	1:1
2.	1:2
3.	1:4
4.	3:2
5.	3:4

FIGURE 5.7.8 Two-story frame.

5.7. Figure 5.7.9 shows a three-hinged frame subjected to horizontal and vertical loads. Select the correct bending moment at node D, M_D, from the following options 1. $M_D = 5\ FL$; 2. $M_D = 10\ FL$; 3. $M_D = 15\ FL$; 4. $M_D = 20\ FL$; 5. $M_D = 25\ FL$.

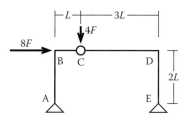

FIGURE 5.7.9 Three-hinged frame.

6 Buckling

6.1 SIMPLE MODELS

In Section 1.7, Chapter 1, we developed an expression to determine the buckling load for a wooden column (Equation 1.7.1). In Chapter 6, we shall derive a similar but more general expression based on what we learned in Chapter 2. In this section, we shall approach the general expression by using simple models of the buckling phenomenon.

Before we do that, we shall conduct a simple test to obtain a feel for the physical phenomenon of buckling.

1. Take a wooden stick approximately 2-ft long with a 1/8-in. (~3.2 mm) square section.
2. Insert the stick into a small block of Styrofoam as shown in Figure 6.1.1a.
3. Attach a binder clip to the top of the stick (Figure 6.1.1b).
4. Place several small magnets onto the binder clip carefully until the stick starts to bend as shown in Figure 6.1.1c. Bending may occur suddenly without warning, or slowly, depending on the straightness of the stick and the arrangement of the weights.

What you have observed is another example of a buckling column albeit on a small scale. Leonhard Euler (1707–1783), a Swiss mathematician and physicist, derived the equations for buckling loads. An example of the "Euler equation" applicable to a cantilever column subjected to an axial load is reproduced in Equation 6.1.1. It should apply to our experiment.

$$P_{cr} = \frac{\pi^2 EI}{4L^2} \approx 2.47 \times \frac{EI}{L^2} \qquad (6.1.1)$$

where E is Young's modulus, I is the moment of inertia, and L is the free length of the stick. Buckling can be very dangerous in a structure. Structural designers need to understand well the buckling mechanism so they can prevent it in structures they design.

Figure 6.1.2a shows a cantilever column subjected to a horizontal load F at its free end. We ignore the self-weight of the column. An expression for the lateral deflection at the free end of such a column was developed in Section 2.8, Chapter 2.

$$v = \frac{FL^3}{3EI} \qquad (2.8.14)$$

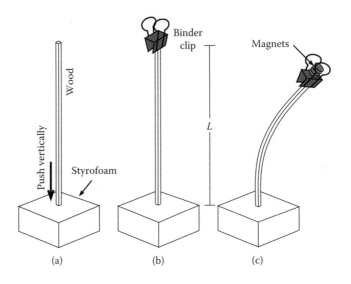

FIGURE 6.1.1 Experiment.

In order to help us understand the buckling phenomenon, we set up an analog column (Figure 6.1.2b). The analog column is rigid throughout its height. It is supported on a pin at the base and its free end is maintained in position by a horizontal, linearly elastic spring attached to the end of the column and a fixed point. In effect, the flexibility of the entire system is concentrated in the spring.

Inspection of Figure 6.1.2a,b will reveal that the spring is analogous to the flexural stiffness of the cantilever column. The cantilever column resists the load F

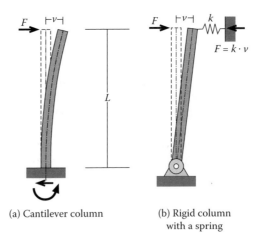

(a) Cantilever column

(b) Rigid column with a spring

FIGURE 6.1.2 System with horizontal spring.

Buckling

because of its flexural stiffness. The analogous column resists the load F with the help of the spring. Without the spring, it would topple over. To improve the analogy, we make the stiffness of the spring, k:

$$k = \frac{3EI}{L^3} \tag{6.1.2}$$

Note that $F = kv$ is the same as that for cantilever column. Even though the lateral stiffness of the analog column is provided by a different mechanism, it mimics the cantilever column successfully.

What would happen if you push the analog column vertically (Figure 6.1.3a)? If you could push the column at its exact cross-sectional center with a load that is exactly vertical, and if the column is perfectly straight and isotropic throughout its length, the column would remain as it is except for a small amount of shortening. However, such a setup is virtually impossible under practical conditions. The vertical is almost always "eccentric" with respect to the resistance axis of the column.

At this time, we stop and introduce the definition of *eccentricity*. It is the perpendicular distance between the axes of column resistance and applied load. We can illustrate it in two dimensions as depicted in Figure 6.1.3a. The axis of column resistance may be represented by the centerline of the column in Figure 6.1.3a. You will note that the axial vertical load P acts at a small distance, e, to the right of the centerline of the column at the point of application. The distance, e, is the eccentricity of the applied vertical load with respect to the center of column resistance.

Because of the eccentricity, the axial load generates a clockwise moment at the top equal to Pe. In response to the applied moment, the analogous column tends to rotate clockwise. This tendency is resisted by the spring. Taking moments about the

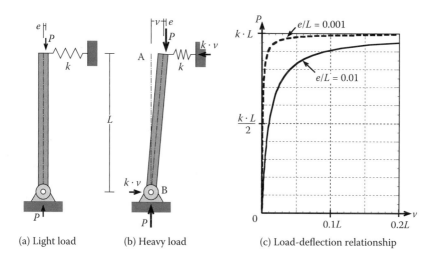

(a) Light load (b) Heavy load (c) Load-deflection relationship

FIGURE 6.1.3 System with horizontal spring.

free end of the analogous column,

$$P \times (v+e) = kv \times L \tag{6.1.3}$$

Rearranging Equation 6.1.3 to define the deflection, v, at the top of the analogous column,

$$v = \frac{Pe}{kL - P} \tag{6.1.4}$$

The solid line and the broken line in Figure 6.1.3c show the relationship between P and v for cases of large eccentricity ($e/L = 0.01$) and small eccentricity ($e/L = 0.001$), respectively. For the case of large eccentricity, the displacement v increases gradually as the load P increases. For the case of small eccentricity, the displacement v increases dramatically when the load P approaches the value of kL because the denominator in Equation 6.1.4 approaches zero. This phenomenon is similar to what we observed when we loaded the wooden stick in the experiment (Figure 6.1.1). We call $P_{cr} = kL$ the buckling load. If we substitute Equation 6.1.2, we obtain

$$P_{cr} = kL = \frac{3EI}{L^2} \tag{6.1.5}$$

This equation is similar to Equation 6.1.1 (the exact solution) except that the coefficient 3 is 20% larger than 2.47. Note that this equation is independent of the *strength* of the spring. The way we have set it up with our assumptions, the spring does not fail. The buckling strength of the analogous column is determined by the *stiffness* of the spring. If we remove the load, the moment $P(v+e)$ will disappear and the spring will push back. The deflection at the top of the analogous column will return to zero. Therefore, we call this failure mechanism "*elastic* buckling" as it is referred to in engineering jargon even though it should be called "buckling in the range of linearly elastic response."

EXERCISE 6.1.1

In GOYA-U1, you can find a system with $k = 0.5$ N/mm and $L = 200$ mm. Fill in Table 6.1.1 for the three cases: $e = 0.1$ mm, $e = 1$ mm, and $e = -1$ mm. Check your results using GOYA-U1.

Figure 6.1.4a shows another simple model with a rotational spring at the bottom of a rigid column. The stiffness of the spring is assumed as

$$K = \frac{3EI}{L^2} \tag{6.1.6}$$

If θ is small enough so that $v = L\sin\theta \approx L\theta$, $F = K\theta/L$ is equivalent to Equation 2.8.14 for the cantilever column. Figure 6.1.4b defines the eccentricity. Figure 6.1.4c shows the deformation caused by the axial force. Moment equilibrium around the spring leads to

$$P \times (L\theta + e) = K\theta \tag{6.1.7}$$

TABLE 6.1.1
Load versus Deflection

Load P (N)	Deflection v (mm)		
	e = 0.1 mm	e = 0.1 mm	e = 0.1 mm
0			
20			
40			
60			
80			
95			

or

$$\theta = \frac{P}{\frac{K}{L} - P} \times \frac{e}{L} \tag{6.1.8}$$

from which

$$P_{cr} = \frac{K}{L} = \frac{3EI}{L^2} \tag{6.1.9}$$

EXERCISE 6.1.2

In GOYA-U2, you can find a system with $K = 20 \times 10^3$ N-mm and $L = 200$ mm. Fill in Table 6.1.2 for the case of $e = 1$ mm in terms of radians and degrees. Check your results using GOYA-U2.

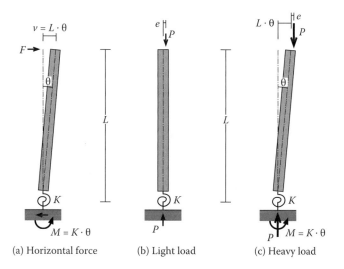

(a) Horizontal force   (b) Light load   (c) Heavy load

FIGURE 6.1.4 System with rotational spring.

TABLE 6.1.2
Load versus Rotation

Load P (N)	Rotation θ (e = 0.1 mm)	
	Unit: Rad.	Unit: Deg.
0		
20		
40		
60		
80		
95		

Now we use a two-spring model that we had considered in Section 2.7, Chapter 2, to evaluate the deflection of a cantilever beam. Figure 6.1.5a shows the model. Recall that each spring represents the flexural deformation of a length of beam equal to $L/2$. The relationship between the bending moments (M_A and M_B) and the rotations of the springs (α_A and α_B in Figure 6.1.5b) are

$$M_A = K\alpha_A \quad \text{and} \quad M_B = K\alpha_B \quad \text{where} \quad K = \frac{2EI}{L} \quad (6.1.10)$$

The free-body diagrams shown in Figure 6.1.5c,d lead to

$$P \times (v_A + e) = M_A = K\alpha_A \quad (6.1.11)$$

$$P \times (v_A + v_B + e) = M_B = K\alpha_B \quad (6.1.12)$$

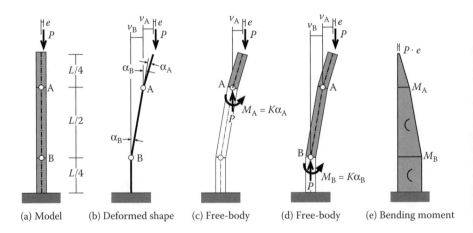

(a) Model (b) Deformed shape (c) Free-body (d) Free-body (e) Bending moment

FIGURE 6.1.5 Two-spring model.

Buckling

Figure 6.1.5e shows the moment distribution along the height. We assume that α_A and α_B are small enough ($\sin \alpha_A \approx \alpha_A$ and $\sin \alpha_B \approx \alpha_B$). From Figure 6.1.5b, we obtain

$$V_A = (\alpha_A + \alpha_B) \times \frac{L}{4} \tag{6.1.13}$$

$$V_B = \alpha_B \times \frac{L}{2} \tag{6.1.14}$$

Substituting Equations 6.1.13 and 6.1.14 into Equations 6.1.11 and 6.1.12,

$$P \times \left[(\alpha_A + \alpha_B) \times \frac{L}{4} + e \right] = K\alpha_A \tag{6.1.15}$$

$$P \times \left[(\alpha_A + \alpha_B) \times \frac{L}{4} + \alpha_B \times \frac{L}{2} + e \right] = K\alpha_B \tag{6.1.16}$$

If we solve these equations in terms of α_A and α_B, we obtain

$$\alpha_A = \frac{(8K - 4PL)Pe}{(PL)^2 - 8KPL + 8K^2} \tag{6.1.17}$$

$$\alpha_B = \frac{8KPe}{(PL)^2 - 8KPL + 8K^2} \tag{6.1.18}$$

Equations 6.1.17 and 6.1.18 have the same denominators. They will be zero if

$$P = \frac{2\sqrt{2}}{\sqrt{2}+1} \times \frac{K}{L} \approx 2.34 \times \frac{EI}{L^2} \tag{6.1.19}$$

or

$$P = \frac{2\sqrt{2}}{\sqrt{2}-1} \times \frac{K}{L} \approx 13.7 \times \frac{EI}{L^2} \tag{6.1.20}$$

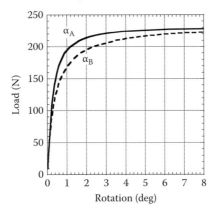

FIGURE 6.1.6 Variation of rotation with load.

TABLE 6.1.3
Load–Moment–Deflection

Load P (N)	Bending Moment		Deflection v (mm)
	M_A (N-m)	M_B (N-m)	
0			
50			
100			
150			
200			
220			

Figure 6.1.6 shows the relationship between the load P and the rotations of the springs (α_A and α_B) assuming $EI = 4 \times 10^6$ N/mm², $L = 200$ mm (or $K = 40{,}000$ N-mm), and $e = 1$ mm. We can see that the rotations increase dramatically as the load approaches $K = 234$ N (the value given by Equation 6.1.19). Note that Equation 6.1.19 is similar to Equation 6.1.1. The error for the approximate solution is only 5%.

EXERCISE 6.1.3

In GOYA-U3, you can find a two-spring model with $L = 200$ mm. Fill in Table 6.1.3 for the case of $EI = 4 \times 10^6$ N/mm² and $e = 1$ mm. Check your results using GOYA-U3.

EXAMPLE 6.1.1

Figure 6.1.7a shows the plan of a 110-story skyscraper built in New York. Each floor is supported by 76 columns with cross sections shown in Figure 6.1.7b. The columns are steel with Young's modulus of 30,000 ksi. The weight of floor per unit area is 200 lbf/ft². The story height is 15 ft (Figure 6.1.7c). Assume that the beams are much stiffer than the columns. Estimate the safety factor of the structure against buckling for the following two cases: (In this application, the safety factor is defined as the buckling strength divided by the axial load.)

Case 1: Structure as it is (the solid line in Figure 6.1.7c).
Case 2: All the beams supporting the second and third floors are destroyed because of fire (the solid line in Figure 6.1.7d). Furthermore, Young's modulus of steel is reduced to 7,500 ksi because of the high temperature.

Hint: Calculate the horizontal stiffness against the force F shown in Figure 6.1.7c,d*. Use the method shown in Figure 6.1.3.

Solution

The total weight of the building is

$$W = 110 \times (200 \text{ ft}^2) \times (200 \text{ lbf/ft}^2) = 8.80 \times 10^8 \text{ lbf}$$

---

* We ignore the possibility of rotation of the building about its vertical axis.

Buckling

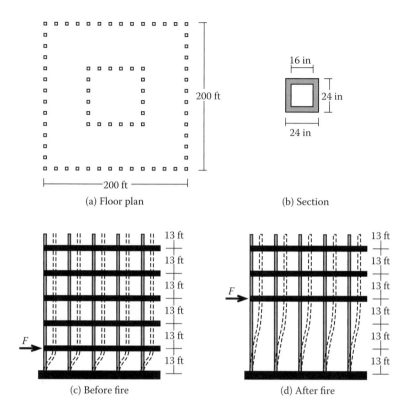

FIGURE 6.1.7 A skyscraper.

The moment of inertia of the column section is

$$I = \frac{24^4}{12} - \frac{16^4}{12} = 22{,}200 \text{ in}^4$$

Recalling Equation 5.5.4 and that there are 76 columns in the first floor, the horizontal stiffness for Case 1 is

$$k = \frac{\Sigma 12EI}{H^3} = \frac{76 \times 12 \times (30{,}000 \times 10^3) \times 22{,}200}{(13 \times 12)^3} = 1.60 \times 10^8 \text{ lbf/in}$$

The critical load for Case 1 is

$$P_{cr} = kH = 1.60 \times 10^8 \times (13 \times 12) = 2.50 \times 10^{10} \text{ lbf}$$

The safety factor is

$$\frac{P_{cr}}{W} = \frac{2.50 \times 10^{10}}{8.80 \times 10^8} \approx 28.4$$

The structure is quite safe against buckling. On the other hand, the horizontal stiffness for Case 2 is

$$k = \frac{\Sigma 12 EI}{H^3} = \frac{76 \times 12 \times (7{,}500 \times 10^3) \times 22{,}200}{(3 \times 13 \times 12)^3} = 1.48 \times 10^6 \text{ lbf/in}$$

The critical load for Case 1 is

$$P_{cr} = kH = 1.48 \times 10^6 \times (3 \times 13 \times 12) = 6.93 \times 10^8 \text{ lbf}$$

The safety factor is

$$\frac{P_{cr}}{W} = \frac{6.93 \times 10^8}{8.80 \times 10^8} = 0.79$$

The structure should collapse as shown by the broken line in Figure 6.1.7d. Collapse is caused by the gravity force, not by a horizontal force. Note that P_{cr} is proportional to EI/H^2. Now that E is 1/4 and H is three times their values before the fire, the safety factor is $(1/4) \times (1/3)^2 = 1/36$ of that in Case 1.

EXAMPLE 6.1.2

A column is loaded in compression as shown in Figure 6.1.8a. Estimate the buckling load assuming that the eccentricity e is small enough. (Hint: the column buckles as shown by the broken lines in Figure 6.1.8a. Use the two-spring model shown in Figure 6.1.8b.)

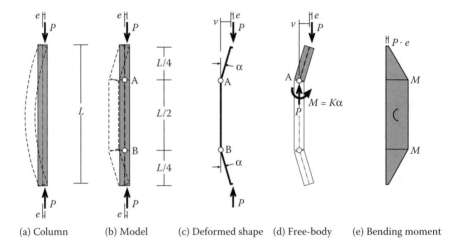

(a) Column   (b) Model   (c) Deformed shape   (d) Free-body   (e) Bending moment

FIGURE 6.1.8 Eccentrically loaded column without moment restraints at its supports.

Buckling

Solution

Figure 6.1.8c shows the deformed shape of the column axis, where α denotes the rotation angle of springs A and B. Figure 6.1.8d shows free-body diagram, which leads to

$$P \times (v + e) = K\alpha$$

Noting $v = \alpha L/4$, we get

$$P \times \left(\frac{\alpha L}{4} + e\right) = K\alpha$$

If we solve this equation in terms of α, we obtain

$$\alpha = \frac{e}{K - \frac{PL}{4}}$$

Assuming the denominator is zero and recalling $K = 2\,EI/L$ (Equation 6.1.10), we conclude that

$$P_{cr} = \frac{4K}{L} = \frac{8EI}{L^2} \qquad (6.1.21)$$

If we assume more than two springs, the coefficient 8 in Equation 6.1.21 will be larger and close to Euler's solution, which will appear in Section 6.2.

$$P_{cr} = \frac{\pi^2 EI}{L^2} \approx \frac{10EI}{L^2} \qquad (6.1.22)$$

Note that the value is four times that in Equation 6.1.1 for a cantilever column.

EXAMPLE 6.1.3

One end of a column is fixed and the other is supported by a vertical roller as shown in Figure 6.1.9a. Estimate the buckling load using the two-spring model shown in Figure 6.1.9b. (Hint: The column will buckle as depicted by the broken line in Figure 6.1.9a,b. Assume an unknown reaction, R, at the roller.)

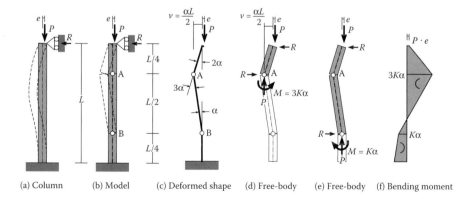

(a) Column (b) Model (c) Deformed shape (d) Free-body (e) Free-body (f) Bending moment

FIGURE 6.1.9 A propped cantilever column.

Solution

Figure 6.1.9c shows the deformed shape of the column axis, where α denotes the rotation angle of spring B. The rotation angle of spring A is 3α. The free-body diagrams shown in Figure 6.1.9d,e lead to

$$P \times \left(\frac{\alpha L}{2} + e\right) - R \times \frac{L}{4} = 3K\alpha$$

$$R \times \left(\frac{L}{4} + \frac{L}{2}\right) - P \times e = K\alpha$$

Eliminating the horizontal reaction R and solving in terms of α, we obtain

$$\alpha = \frac{4Pe}{20K - 3PL}$$

Assuming the denominator is zero and recalling $K = 2\, EI/L$ we conclude that

$$P_{cr} = \frac{20K}{3L} = \frac{40EI}{3L^2} \tag{6.1.23}$$

If we assume more than two springs, the coefficient 40/3 in Equation 6.1.23 will be larger and close to Euler's solution.

$$P_{cr} = \frac{2\pi^2 EI}{L^2} \approx \frac{20EI}{L^2} \tag{6.1.24}$$

Note that the value is eight times that in Equation 6.1.1 for a cantilever column.*

6.2 CONTINUOUSLY DEFORMABLE MODEL

Having dealt with the problem of buckling using simple and rigid-discrete models, we are ready to derive Euler's equation in reference to a continuously deformable model. To simplify the notation, we shall refer to a horizontal member (Figure 6.2.1a) or a cantilever column rotated through 90°. In Figure 6.2.1b, v_0 denotes the deflection at the free end. Figure 6.2.1c shows the deflected shape of the member axis, where v denotes the deflection of the member at a distance x from the fixed end. The free-body diagram shown in Figure 6.2.1d determines the bending moment at any section with a known deflection v.

$$M = P \times (e + v_0 - v) \tag{6.2.1}$$

Figure 6.2.1e shows the bending-moment diagram corresponding to Equation 6.2.1.
Recall the following equation derived in Section 2.8, Chapter 2, to describe the relationship between curvature and bending moment.

$$\frac{d^2 v}{dx^2} = \frac{M}{EI} \tag{2.8.9}$$

---

* Leonhard Euler (1707–1783) was completely blind for the last seventeen years of his life, during which time he produced almost half of his total work output. He had extraordinary powers of memory and mental calculation.

Buckling

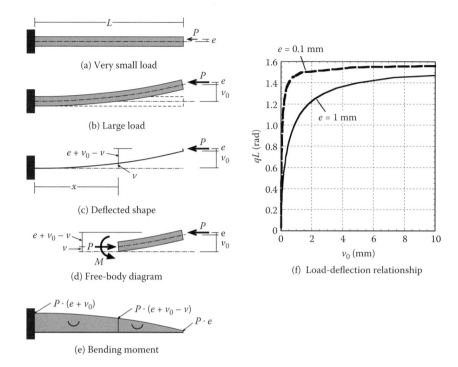

FIGURE 6.2.1 Buckling of cantilever beam.

Replacing M as defined by Equation 6.2.1, we obtain

$$\frac{d^2v}{dx^2} = \frac{P \times (e + v_0 - v)}{EI} \quad (6.2.2)$$

If we let $q = \sqrt{P/EI}$,

$$\frac{d^2v}{dx^2} + q^2 v = q^2 (e + v_0) \quad (6.2.3)$$

The solution of the differential equation is of the form

$$v = C_1 \sin qx + C_2 \cos qx + e + v_0 \quad (6.2.4)$$

where C_1 and C_2 are constants of integration*.

---

* The second derivative of Equation 6.2.4 is

$$\frac{d^2v}{dx^2} = -C_1 q^2 \sin qx - C_2 q^2 \cos qx \quad (6.2.5)$$

If we substitute Equations 6.2.4 and 6.2.5 into Equation 6.2.3, we find that Equation 6.2.3 is satisfied. Therefore, we conclude that Equation 6.2.4 is the solution.

To evaluate the constants C_1 and C_2, we use the boundary conditions $v = 0$ and $dv/dx = 0$ at $x = 0$ (at the fixed end), from which we get $C_1 = 0$ and $C_2 = -(e + v_0)$. Substituting these values into Equation 6.2.4, we obtain

$$v = (e + v_0)(1 - \cos qx) \tag{6.2.6}$$

To determine the deflection at the free end, v_0, we use the other boundary condition: $v = v_0$ at $x = L$ (at the free end). This leads to

$$v_0 = e\left(\frac{1}{\cos qL} - 1\right) \tag{6.2.7}$$

Figure 6.2.1f shows the relationship between qL and v_0 for the case of $L = 100$ mm. The value qL is indicated in terms of radians (no physical unit) because the unit of $q = \sqrt{P/EI}$ is a reciprocal of length. Note that v_0 increases to infinity as qL approaches $\pi/2 \approx 1.57$ even if the eccentricity e is very small. The reason for this result is that $\cos qL$ in Equation 6.2.7 approaches zero. Substituting $q = \sqrt{P/EI}$ into $qL = \pi/2$ and solving for P, we get Euler's equation for the buckling load of a cantilever column.

$$P_{cr} = \frac{\pi^2 EI}{4L^2} \tag{6.2.8}$$

EXERCISE

In GOYA-U4, you can find the case illustrated in Figure 6.2.2a: a 200-mm long column with a cross section of 10 mm × 15 mm and a Young's modulus of 1000 N/mm².

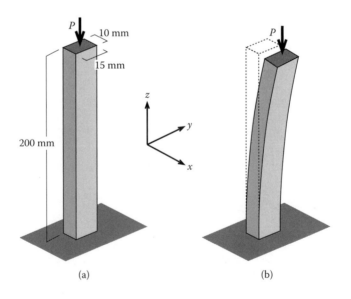

(a)　　　　　　　　　　(b)

FIGURE 6.2.2 Column.

Buckling

TABLE 6.2.1
Results for Eccentricities of 0.1 and 1mm

Load P (N)	q (/mm)	qL		cosqL	v_0 (mm)	
		Unit: Rad.	Unit: Deg.		e = 1 mm	e = 0.1 mm
0						
25						
50						
75						

Because the moment of inertia is smaller around the y-axis, the column deflects in direction x as shown in Figure 6.2.2b. Calculate the moment of inertia around the y-axis and calculate the buckling load. Fill in Table 6.2.1 for the two cases: $e = 0.1$ mm and $e = 1$ mm. Check your results using GOYA-U4.

EXAMPLE 6.2.1

Take a spaghetti strand 200 mm (~8 in.) long with a diameter of 1.6 mm (~0.06 in.). Assume that Young's modulus is 2000 N/mm², the compressive strength is 200 N/mm², and the tensile strength is 20 N/mm². Compute the buckling loads for the spaghetti strand for the three types of loading shown in Figure 6.2.3. Check your results in the kitchen using a scale.

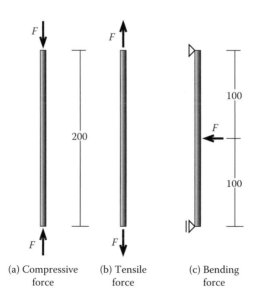

(a) Compressive force (b) Tensile force (c) Bending force

FIGURE 6.2.3 Loading of spaghetti.

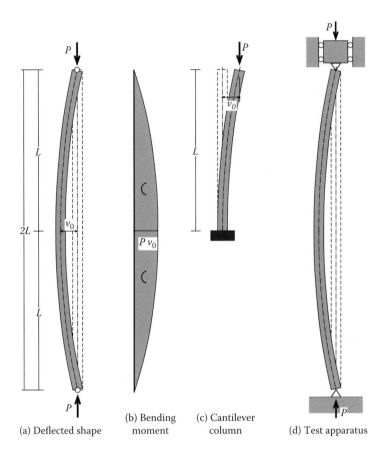

(a) Deflected shape (b) Bending moment (c) Cantilever column (d) Test apparatus

FIGURE 6.2.4 Pin-Supported column.

Solution

Moment of inertia of the spaghetti strand is

$$I = \frac{\pi r^4}{4} = \frac{3.14 \times 0.8^2}{4} = 0.32 \text{ mm}^4$$

Figure 6.2.4a shows the deformed shape of the strand, with length $2L$, after buckling. Figure 6.2.4b shows the bending-moment diagram, where eccentricity is assumed to be negligible. The condition is equivalent to that of a cantilever column of length L shown in Figure 6.2.4c. Therefore, the buckling load is

$$P_{cr} = \frac{\pi^2 EI}{4L^2} = \frac{3.14^2 \times 2000 \times 0.322}{4 \times 100^2} = 0.16 \text{ N} \qquad \text{(or 0.6 oz)}$$

The cross-sectional area of the spaghetti strand is

$$A = \pi r^2 = 3.14 \times 0.8^2 = 2.0 \text{ mm}^2$$

Buckling

The force required to break the strand by pulling is the product of the cross-sectional area and the tensile strength.

$$F = A\sigma = 2.0 \times 20 = 40 \text{ N} \qquad \text{(or 8.9 lb)}$$

Section modulus of the spaghetti is

$$Z = \frac{I}{r} = \frac{\pi r^3}{4} = \frac{3.14 \times 0.8^3}{4} = 0.4 \text{ mm}^3$$

Flexural strength is determined using the bending strength ($M = Z\sigma$) and the span length L.

$$F = \frac{2M}{L} = \frac{2Z\sigma}{L} = \frac{2 \times 0.4 \times 20}{100} = 0.16 \text{ N} \qquad \text{(or 0.6 oz)}$$

If you test a 100-mm-long spaghetti, the flexural strength will be double whereas the buckling strength will be four times. In GOYA-U5, you can have a test apparatus shown in Figure 6.2.4d. We use such an apparatus when we conduct a test of buckling of an I-shaped steel column.

Isaac Newton (1643–1727) showed that the laws of physics observed on Earth are also observed in space. You should understand that the laws of mechanics observed in the kitchen using a spaghetti strand are also observed in skyscrapers that use steel columns.

EXAMPLE 6.2.2

If you keep loading the spaghetti strand in Example 6.2.1 after buckling, it will break. Estimate the deflection v_0 at fracture. (Hint: The spaghetti strand will break if the maximum tensile stress reaches its tensile strength. Look in GOYA-U5 and see the stress distribution in the section at midspan.)

Solution

Figure 6.2.5a shows the stress distribution of a column subjected to pure compression, where P is assumed positive though in compression. If we apply a small bending moment, the stress distribution will be trapezoidal as shown in Figure 6.2.5b (Section 5.1, Chapter 5). If we increase the bending moment, the stress distribution will be as shown in Figure 6.2.5c. If the maximum tensile stress reaches the strength, the column will fail.

$$\sigma = -\frac{P}{A} + \frac{M}{Z} \qquad (6.2.9)$$

Substituting $M = Pv_0$ and $\sigma = 20$ N/mm^2 into Equation 6.2.9, and solving for v_0, we get

$$v_0 = \left(\sigma + \frac{P}{A}\right) \times \frac{Z}{P} = \left(20 + \frac{0.159}{2.01}\right) \times \frac{0.402}{0.159} = 50.8 \text{ mm} \qquad \text{(or 2.0 in.)}$$

The maximum compressive stress is

$$\sigma = -\frac{P}{A} - \frac{M}{Z} = -\frac{P}{A} - \frac{Pv_0}{Z} = -\frac{0.159}{2.01} - \frac{0.159 \times 50.8}{0.402} = -0.08 - 20.1 = -20.2 \text{ N/mm}^2$$

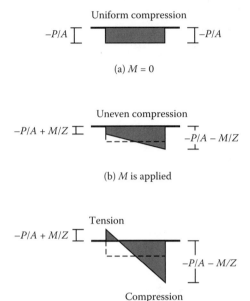

FIGURE 6.2.5 Stress distribution.

Note that the magnitude is almost equal to the maximum tensile stress, $\sigma = 20$ N/mm^2. The stress distribution in a buckled column is quite similar to that in a beam without axial force.

EXERCISE 6.2.1

Compute the deflection v_0 at fracture for a 100-mm-long spaghetti strand. The answer will be about 1/4 of that obtained in Example 6.2.2.

EXERCISE 6.2.2

Compute the deflection of the spaghetti strand v_0 at fracture under the loading shown in Figure 6.2.6 for cases of $L = 100$ mm and 50 mm. The answers will be smaller than those for buckling, but not very different, because the bending-moment distributions are similar.

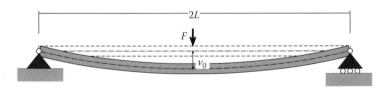

FIGURE 6.2.6 Simply supported beam.

Buckling

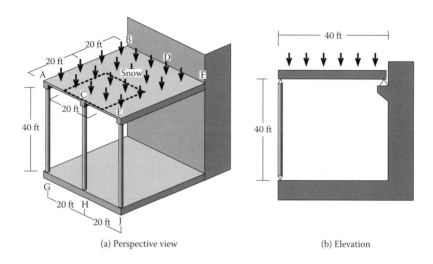

(a) Perspective view (b) Elevation

FIGURE 6.3.1 Canopy in snowy region.

6.3 PROBLEMS

6.1 Figure 6.3.1a shows a canopy that extends out from a building. We may assume that the building is stiff and strong enough not to fail under the conditions specified. The canopy is supported by three beams: AB, CD, and EF. You may assume the beams to be simply supported (Figure 6.3.1b). Each beam is supported by a steel pipe-column having an outside diameter of 8.0 in. and an inside diameter of 6.0 in. The connections at both ends of each column may be assumed to be pins. Assume that Young's modulus for steel is 30,000 ksi. The building is in a snowy region, and the maximum possible snow load per unit area of the canopy is estimated to be 1,000 lbf/ft². Assume that the self weights of the canopy and the beams are negligible, and the tributary area* of column CH is $20 \times 20 = 400$ ft² as indicated by the broken lines in Figure 6.3.1a. Check if column CH is safe against buckling.

6.2 Assume that we make the connections between the foundation beams and the columns continuous so that the columns may be assumed to be fixed at their bases. Check if the columns are safe against buckling under the specified snow load.

6.3 Assume that we have a rigid beam between A and E. Check if the columns are safe against buckling under the specified snow load.

---

* The "tributary area" for a particular column is defined as the area that contributes load to that column. In this case, we assume that the moment restraints on the slab across lines AB, CD, and EF are similar. Accordingly, half the load on the slab goes to beam CD. Because beam CD is supported simply, half of the load it carries goes to support C. The tributary area is one-fourth of the area ABEF or $20 \times 20 = 400$ ft².

Answers to Problems

1.1 Equilibrium of forces at node A is shown graphically in Figure 1.1. Because $5\sqrt{3} \approx 8.7$, No. 5 is correct.

1.2 Noting that (stress) = (axial force)/(cross-sectional area), the stresses of members AB and AC are $\sigma_{AB} = 5\sqrt{3}/100$ and $\sigma_{AC} = 5/100$ (kN/mm²), respectively. Because (strain) = (stress)/(Young's modulus), strains in members AB and AC are $\varepsilon_{AB} = 5\sqrt{3}/(100 \times 200)$ and $\varepsilon_{AC} = 5/(100 \times 200)$, respectively. Recalling (deformation) = (strain) × (length), we have

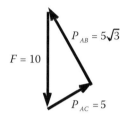

FIGURE 1.1 Equilibrium.

$$e_{AB} = \frac{5\sqrt{3}}{100 \times 200} \times 10,000 = \frac{5\sqrt{3}}{2} \approx 4.3 \text{ (mm)}$$

$$e_{AC} = \frac{5}{100 \times 200} \times 10,000\sqrt{3} = \frac{5\sqrt{3}}{2} \approx 4.3 \text{ (mm)}$$

No. 3 is correct.

1.3 Because the strength of the material is 200 N/mm² and its cross-sectional area is 100 mm², the maximum possible axial force is $200 \times 100 = 20$ kN. On the other hand, the axial forces in members AB and AC caused by an external force F are as follows (see Figure 1.1).

$$P_{AB} = \frac{\sqrt{3}}{2} F \qquad P_{AC} = \frac{1}{2} F$$

Noting that P_{AB} is larger than P_{AC}, we can assume that the truss fails if $P_{AB} = 20$ kN.

$$P_{AB} = \frac{\sqrt{3}}{2} F = 20 \text{ leads to } F = \frac{2}{\sqrt{3}} \times 20 \approx 23.1 \text{ (kN)}$$

No. 4 is correct.

1.4 Equilibrium of the moments around support A leads to a reaction at support E expressed as

$$R_E = \frac{5}{20} \times 20 = 5 \text{ (N)}.$$

Equilibrium of vertical forces leads to

$$R_A = 20 - R_E = 15 \text{ (N)}.$$

Equilibrium at node A shown in Figure 1.2a yields $P_{AC} = 15$ kN, while equilibrium at node E shown in Figure 1.2b yields $P_{CE} = 5$ kN.
No. 1 is correct.

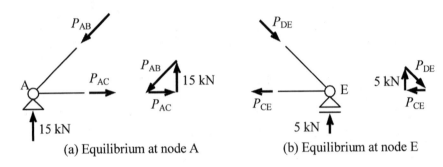

(a) Equilibrium at node A   (b) Equilibrium at node E

FIGURE 1.2 Equilibrium.

1.5 The elongations of members AC and CE are calculated as below.

$$e_{AC} = \frac{15}{100 \times 200} \times 10{,}000 = 7.5 \text{ (mm)}$$

$$e_{CE} = \frac{5}{100 \times 200} \times 10{,}000 = 2.5 \text{ (mm)}$$

Because node A does not move horizontally, node E moves right by $e_{AC} + e_{CE} = 10$ mm.
No. 5 is correct.

1.6 Because of symmetry, each reaction is 0.5 F. Equilibrium at node A is shown in Figure 1.3. It leads to

$$P_{AB} = -\frac{0.5F}{\sin\theta} \text{ (compression)} \quad \text{and} \quad P_{AC} = \frac{0.5F}{\tan\theta} \text{ (tension)}.$$

Next, equilibrium at node C (Figure 1.4) leads to

$$P_{AC} = P_{CE} \text{ (tension)} \quad \text{and} \quad P_{BC} = 0.$$

Because of symmetry, we have $P_{FG} = 0$. These results indicate that No. 4 is correct.

Answers to Problems

FIGURE 1.3 Equilibrium at node A.

FIGURE 1.4 Equilibrium at node C.

Let us check the other axial forces. Equilibrium at the node B is shown in Figure 1.5. Because the direction of P_{AB} is same as that of P_{BD} and $P_{BC} = 0$, we have

$$P_{AB} = P_{BD} \text{ (compression)} \quad \text{and} \quad P_{BE} = 0.$$

Note that if $P_{BE} \neq 0$ the node is not in equilibrium. Because of symmetry, we have $P_{EF} = 0$. Equilibrium at node E shown in Figure 1.6 leads to

$$P_{CE} = P_{EG} \text{ (tension)} \quad \text{and} \quad P_{DE} = 0.$$

FIGURE 1.5 Equilibrium at node B.

FIGURE 1.6 Equilibrium at node E.

These results naturally satisfy equilibrium at node D shown in Figure 1.7.

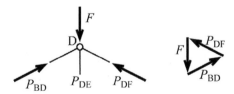

FIGURE 1.7 Equilibrium at node D.

To summarize the results above, we note that the members between nodes A-D, D-H and A-H have axial forces and the other members do not.

In this sense, this truss (Figure 1.8a) is equivalent to the simple truss shown in Figure 1.8b. The horizontal and vertical displacements of node D (u_{Dx} and u_{Dy}) are also the same. In the truss shown in Figure 1.8a, nodes C, E and G move down because members BC, BE, DE, EF and FG do not elongate or shorten. You may think the horizontal displacement of node H (u_{Hx}) in Figure 1.8a should be smaller than that in Figure 1.8b because the bottom chord in Figure 1.8a deflects downward. This is true, but the difference between them is negligible.

1.7 Because of symmetry, the reactions are 1.5 F. Equilibrium at node A is shown in Figure 1.9. It leads to

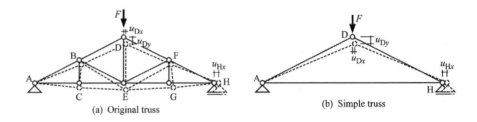

FIGURE 1.8 Two trusses.

$$P_{AB} = -\frac{1.5F}{\sin\theta} \text{ (compression)} \quad \text{and} \quad P_{AC} = \frac{1.5F}{\tan\theta} \text{ (tension)}.$$

Next, equilibrium at node C shown in Figure 1.10 leads to

$$P_{AC} = P_{CE} \text{ (tension)} \quad \text{and} \quad P_{BC} = 0.$$

Because of symmetry, we have $P_{FG} = 0$. These results indicate that No. 4 is correct.

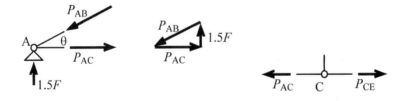

FIGURE 1.9 Equilibrium at node A.   **FIGURE 1.10** Equilibrium at node C.

Answers to Problems

Let us check the other axial forces. Considering equilibrium at node B shown in Figure 1.11 in the x- and y-directions, we have the following equations.

X-direction: $P_{AB}\cos\theta - P_{BD}\cos\theta - P_{BE}\cos\theta = 0$

Y-direction: $P_{AB}\sin\theta - F - P_{BD}\sin\theta + P_{BE}\sin\theta = 0$

Therefore,

$$P_{BD} = -\frac{F}{\sin\theta} \text{ (compression)} \quad \text{and} \quad P_{BE} = -\frac{0.5F}{\sin\theta} \text{ (compression)}.$$

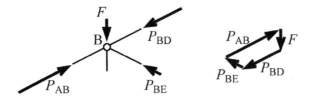

FIGURE 1.11 Equilibrium at node B.

Furthermore, equilibrium at node E (Figure 1.12) in y-direction leads to the following equation.

$$P_{DE} = -2P_{BE}\sin\theta = F \text{ (tension)}.$$

Why is member DE in tension in spite of the downward force at node D? This puzzle may be solved if you consider equilibrium at node D (Figure 1.13). Because the compressive forces in members BD and DF are so large, member DE is in tension.

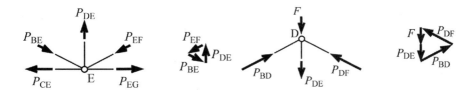

FIGURE 1.12 Equilibrium at node E. **FIGURE 1.13** Equilibrium at node D.

Figure 1.14 shows the result of the truss by analysis using GOYA-A. Node E moves downward slightly more than node D, which indicates that member DE is in tension.

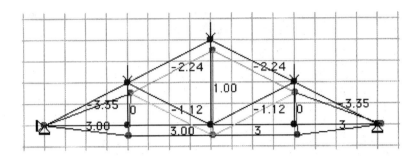

FIGURE 1.14 Result by GOYA-A.

1.8 Because of symmetry, each reaction is 0.5 F. Considering equilibrium at node C as shown in Figure 1.15 leads to

$$P_{AC} = -\frac{0.5F}{\sin 60} = -\frac{F}{\sqrt{3}} \text{ (compression).}$$

Also, equilibrium at node A is shown in Figure 1.16 leading to

$$P_{AB} = P_{AC} = -\frac{F}{\sqrt{3}} \text{ (compression).}$$

No. 4 is correct.

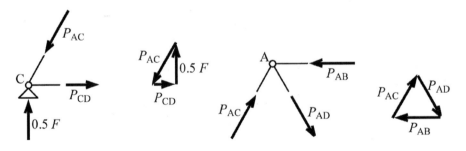

FIGURE 1.15 Equilibrium at node C.   **FIGURE 1.16** Equilibrium at node A.

1.9 Because of symmetry, each reaction is 2.5 F. The smartest approach to determine the axial force in member AB is to cut the truss as shown in Figure 1.17 and consider equilibrium of moment around node C. Denoting the clockwise moment as positive, we have

$$\sum M = 2.5F \times 4 - F \times 4 - F \times 2 - P_{AB} \times 2 = 0,$$

which leads to $P_{AB} = 2F$ (tension).
No. 1 is correct.

Answers to Problems 317

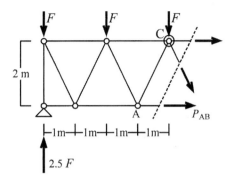

FIGURE 1.17 Free-body diagram.

1.10 Because of symmetry, each reaction is F as shown in Figure 1.18.

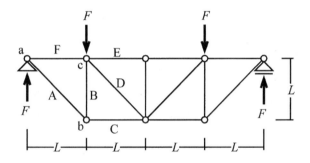

FIGURE 1.18 Reactions.

Equilibrium at node a shown in Figure 1.19a leads to

$$P_A = +\frac{F}{\sin 45} = +\sqrt{2}F \quad \text{(tension) and}$$

$$P_F = -F \quad \text{(compression)}.$$

Also, equilibrium at node b in Figure 1.19b leads to

$$P_B = -F \quad \text{(compression) and}$$

$$P_C = +F \quad \text{(tension)}.$$

Finally, equilibrium at node c in Figure 1.19c leads to

$$P_D = 0 \quad \text{and}$$

$$P_E = -F \quad \text{(compression)}.$$

No. 4 is incorrect.

Sketch how the truss deforms. Then use GOYA-A to determine the deflected shape. You will find that the center of the truss moves down more than the loaded nodes do because member E shortens (it is in compression) and member C elongates (it is in tension).

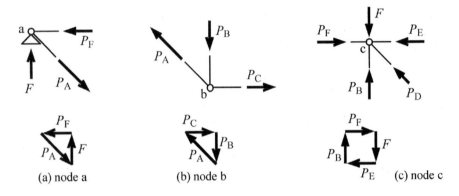

FIGURE 1.19 Equilibrium.

You can also calculate the axial forces in members C, D and E considering the free body shown in Figure 1.20. Equilibrium in the vertical direction requires $P_D = 0$. Equilibrium of moment around the support leads to $P_C = +F$. Equilibrium in the horizontal direction requires $P_E = -F$.

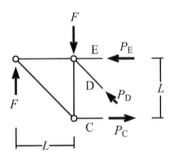

FIGURE 1.20 Free body diagram.

1.11 Because of symmetry, the reactions are 2.5F as shown in Figure 1.21a. The wisest approach to finding the axial force in member AB is to cut the truss as shown in Figure 1.21b and consider equilibrium of moment around node C. Denoting the clockwise moment as positive, we have

$$\sum M = F \times L + F \times 2L - 2.5F \times L - P_{AB} \times L = 0,$$

which leads to $P_{AB} = +0.5F$ (tension). Thus, No. 3 is correct. Imagine how the truss deforms, then determine the deflected shape using GOYA-A. You will find that the center of the truss moves upward because member AB elongates (it is in tension) while member DC shortens (it is in compression).

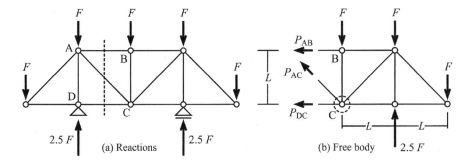

FIGURE 1.21 Reactions and free body diagram.

1.12 Note that roller support D has a vertical reaction only as shown in Figure 1.22a. Equilibrium of moments around node E indicates the reaction at D to be $(2/5)F$. The reaction at E should be $F - (2/5)F = (3/5)F$. The smartest approach to determine the axial force in member AB is to cut the truss as shown in Figure 1.22b and consider moment equilibrium around node C. Denoting the clockwise moment as positive, we have

$$\sum M = \frac{2}{5}F \times 3L - P_{AB} \times \sqrt{3}L = 0,$$

which leads to $P_{AB} = \frac{6}{5\sqrt{3}} F$ (tension).
No. 2 is correct.

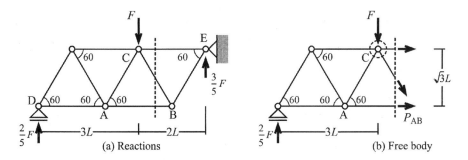

FIGURE 1.22 Reactions and free body diagram.

1.13 Note that roller support E in Figure 1.23a can develop only a horizontal reaction as shown. Equilibrium of moments around node D leads to the reaction at E as $\sqrt{3}F$. Equilibrium of forces in the horizontal and vertical directions require reactions at support D as shown in Figure 1.23a. The smartest approach to determine the axial force in member AB is to cut the truss as shown in Figure 1.23b and consider equilibrium of moments around node C. Denoting the clockwise moment as positive, we have

$$\sum M = F \times 3L - \sqrt{3}F \times \sqrt{3}L - P_{AB} \times \sqrt{3}L = 0,$$

leading to $P_{AB} = 0$.
No. 3 is correct.

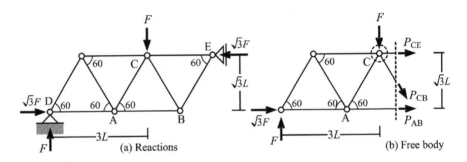

FIGURE 1.23 Reactions and free body diagram.

Incidentally, equilibrium at node E leads to $P_{BE} = 0$ and equilibrium at node B leads to $P_{CB} = 0$. If we remove the members of zero axial force (AB, BE and CB), however, the truss becomes unstable as shown in Figure 1.24.

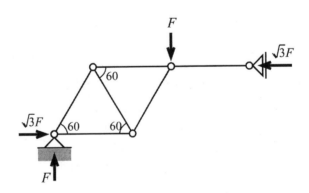

FIGURE 1.24 Unstable truss.

Figure 1.25 compares the results from GOYA-A of the trusses in the problems above, where F is assumed to be 50 N. We note that both the axial forces and the deformations are smaller in the truss described in problem 1.12 than that in problem 1.13.

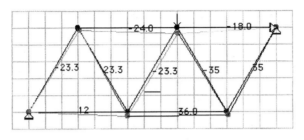

(a) Truss (problem 1-12)

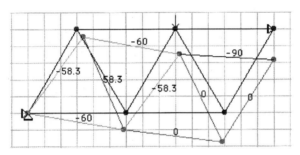

(b) Truss (problem 1.13)

FIGURE 1.25 Results by GOYA-A.

1.14 First, let us evaluate the reactions. Equilibrium of forces in the horizontal direction requires that the horizontal reaction at support D is zero (Figure 1.26). Next, equilibrium of moments around node C requires:

$$\sum M = F \times L + 2F \times 2L + 3F \times 3L - R_{Dy} \times 4L = 0,$$

leading to $R_{Dy} = 3.5F$. The smartest approach to determine the axial force in member AB is to cut the truss as shown in Figure 1.26 and consider equilibrium of forces in the vertical direction. Denoting forces up as positive, we have

$$\sum Y = -3F + 3.5F - T\cos 45 = 0,$$

leading to $T = +\frac{1}{\sqrt{2}} F$ (tension).
No. 4 is correct.

Did you assume AB to be in compression because of the force (3F) at node B? No, the effect of the load at the center (2F) is larger.

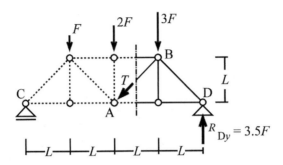

FIGURE 1.26 Free body diagram.

2.1 Wherever you cut the beam and examine the free-body diagram, you will have the same bending moment M acting on the section. You conclude that the moment distribution is uniform and the correct answer may be 4 or 5. Because the moment is uniform and the section of the beam does not change (prismatic beam), the curvature is also uniform. Therefore, the deflection should not be zero. Choice 5 must be correct.
Let us calculate the deflection just to make sure. Integrating

$$\frac{d^2v}{dx^2} = \frac{M}{EI}$$

with the boundary condition at the fixed end ($dv/dx = 0$ and $v = 0$ at $x = 0$), we get

$$v = \frac{M}{2EI} x^2$$

The deflection listed in 5 is correct.

2.2 We start with the knowledge, from equilibrium of the entire structure (Figure 2.1a), that the total force in the reactions at ends A and C must be equal and opposite to the applied load.

$$F = R_A + R_C$$

Recall Equation 2.6.17, which gives the deflection of a cantilever beam under a concentrated load at its free end (this equation is very useful and worth memorizing). Applying this equation to the beam AB, we determine its free-end deflection to be

$$v = \frac{R_A L^3}{3EI}$$

On the other hand, applying this equation to beam BC gives

$$v = \frac{R_C(2L)^3}{3EI} = \frac{8R_C L^3}{3EI}$$

Observing that the hinge forces these deflections to be the same, we have

$$\frac{R_A}{R_C} = 8$$

The correct answer is 5. The reactions and deformation are shown in Figure 2.1b. Figure 2.1c shows the bending moment distribution, where we should note that the bending moment in beam BC is much smaller than that in beam AB.

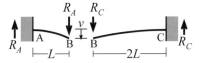

FIGURE 2.1 Connected cantilever beams.

2.3 As we have seen in Example 2.8.5, the deflection of a beam subjected to a uniformly distributed load is proportional to the fourth power of its length. Therefore, the deflection of Beam A would be $2^4 = 16$ times that of Beam B, if the two beams had the same section. Deflection for a given loading is inversely proportional to the moment of inertia of the beam section. The moment of inertia of Beam B is $2^3 = 8$ times that of Beam A. which indicates that its deflection should be (1/8) times that of beam A. Therefore, $\delta_B/\delta_A = 16/8 = 2$. The correct answer is 2.

2.4 The cantilever beam in the problem is equivalent to the beam in Figure 2.2(a), where force R is applied to prevent deflection at the free end. Problem 2.4 is, therefore, identical to Example 2.8.4, where deflections caused by two loads are superimposed. In problem 2.4, however, we can determine the correct answer more easily. First, note that the deformation shall be downward concave near the fixed end and concave upward near mid-span and the free end as shown in Figure 2.2a. Among the five options, the first and the second are clearly wrong because the bending moments are all positive, which would make the free end deflect upward. The third option is also wrong because the bending moment at the free end is not zero: it should be zero because there is no couple at the free end. To examine the remaining two options, let us examine the shear force diagrams corresponding to the fourth and fifth options based on $V = dM/dx$. The fourth one (Figure 2.2b) indicates that the force at mid-span should be $8 + 5 = 13$ N; so, it is wrong. The fifth one (Figure 2.2c) indicates the force is $11 + 5 = 16$ N. The correct answer is 5.

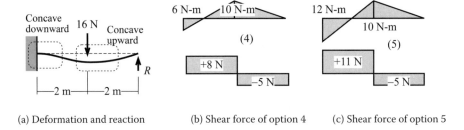

(a) Deformation and reaction (b) Shear force of option 4 (c) Shear force of option 5

FIGURE 2.2 Propped cantilever beam with a load at mid-span.

2.5 The cantilever beam in the problem is equivalent to the beam in Figure 2.3a, where the force R is applied so that the deflection of the free end is zero. Note that the force R should be downward because the couple of 10 N-m would induce an upward deflection without it. As a result, the beam should deform concave downward near the fixed end and concave upward near mid-span and the free end as shown in Figure 2.3a. Among the five options, the first and the second are clearly wrong because the bending moments are all positive. Figure 2.3b shows the deformation caused by the moment distribution for option (4) leading us to conclude that this option is wrong. Note that this moment distribution leads to a symmetrical deformation with respect to mid-span, which would make the free end deflect downward. The concave upward deformation should be larger than the concave downward deformation in order to let the free end stay at its original position. Option (5) is worse than (4) because this moment distribution makes the free end deflect downward even more. The correct answer should be (3). As we can see from the slope of the moment distribution dM/dx, the reaction at the free end is $R = 15/4$ N. We can also obtain this value equating the results of Examples 2.8.1 and 2.8.2:

$$v = \frac{ML^2}{2EI} \quad \text{and} \quad v = \frac{RL^3}{3EI} \quad \text{which lead to} \quad R = \frac{3M}{2L} = \frac{3 \times 10}{2 \times 4} = \frac{15}{4} \text{ N.}$$

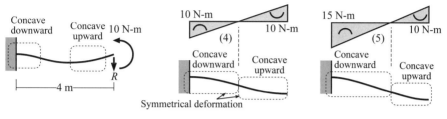

(a) Deformation and reaction (b) Deformation of option 4 (c) Deformation of option 5

FIGURE 2.3 Propped cantilever beam with a couple.

2.6 First, let us check whether the options satisfy the uniform load and shear, $dV/dx = w$. The first one is incorrect because $dV/dx = 0$. The third one is also wrong because $dV/dx = -12/4 = -3$. The remaining three options satisfy $dV/dx = -4$.

Next, let us draw the bending moment diagram integrating $dM/dx = V$ and noting $M = 0$ at the free end as shown in the middle row of Figure 2.4, where the broken lines indicate the location of $V = dM/dx = 0$. Now, we can visualize the deflection as shown in the bottom row of Figure 2.4. Option (2) is wrong because the beam deformation is concave downward at all points in the beam span. Option (5) is also wrong because the beam deformation is concave upward at all points. The correct answer should be the remaining option (4). The shear force distribution for this option indicates that the reaction at the free end is $R = 6$ N. We can also obtain this value equating the results of Examples 2.8.2 and 2.8.5:

$$v = \frac{RL^3}{3EI} \quad \text{and} \quad v = \frac{wL^4}{8EI} \quad \text{which lead to} \quad R = \frac{3wL}{8} = \frac{3 \times 4 \times 4}{8} = 6 \text{ N.}$$

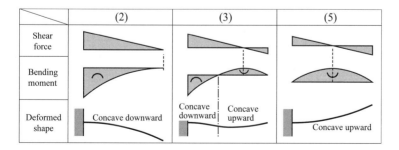

FIGURE 2.4 Propped cantilever beam with a distributed load.

2.7 The moments of inertia of the sections are obtained from $I = bh^3/12$.

$$I_A = \frac{1}{12} \cdot D \cdot D^3 = \frac{1}{12} D^4$$

$$I_B = \frac{1}{12} \cdot 2D \cdot (2D)^3 = 2^4 \times \frac{1}{12} D^4 = 2^4 \times I_A$$

Denoting the length of the columns as L and the horizontal forces on the columns as F_A and F_B, the lateral deflections of the tops of the columns are

$$v_A = \frac{F_A L^3}{3EI_A} \quad \text{and} \quad v_B = \frac{F_B L^3}{3EI_B}$$

Because these deflections should be the same ($v_A = v_B$), we have

$$F_B = \frac{I_B}{I_A} \times F_A = 2^4 \times F_A$$

The bending moments at the fixed ends of the columns are

$$M_A = F_A L \quad \text{and} \quad M_B = F_B L = 2^4 \times M_A$$

The section moduli are obtained from $Z = bh^2/6$.

$$Z_A = \frac{1}{6} \cdot D \cdot D^2 = \frac{1}{6} D^3$$

$$Z_B = \frac{1}{6} \cdot 2D \cdot (2D)^2 = 2^3 \times \frac{1}{6} D^3 = 2^3 \times Z_a$$

The stresses at points a and b are obtained from Equation 2.5.7.

$$\sigma_a = \frac{M_A}{Z_A} \quad \text{and} \quad \sigma_b = \frac{M_B}{Z_B} = \frac{2^4 \times M_A}{2^3 \times Z_A} = 2\sigma_a$$

The correct answer is 2.

2.8 We take the x-coordinate from the free end as shown in Figure 2.5 and denote the width and the depth at the fixed end as b and h. Then, we express the cross-sectional area of the beam as follows.

$$A = \frac{x}{L} \cdot bh$$

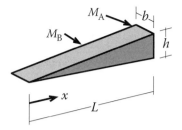

FIGURE 2.5 Cantilever beam with linearly varying depth.

We define the gravity force per volume as ρg. The gravity force per unit length is then

$$w = -A \times \rho g = -\frac{x}{L} \cdot bh\rho g$$

The negative sign indicates that the gravity force is downward. Integrating this with the boundary condition ($V = 0$ at $x = 0$), we have the shear force:

$$V = -\frac{x^2}{2L} \cdot bh\rho g$$

Integrating again with the boundary condition ($M = 0$ at $x = 0$), we get the bending moment:

$$M = -\frac{x^3}{6L} \cdot bh\rho g$$

The bending moments at the fixed end ($x = L$) and at mid-span ($x = L/2$) are

$$M_A = -\frac{L^2}{6} \cdot bh\rho g \quad \text{and} \quad M_B = -\frac{L^2}{48} \cdot bh\rho g$$

Therefore, the correct answer is 4.

We can solve this problem using another approach. Figure 2.6a shows the equilibrium with the beam cut at the fixed end. We substitute a concentrated load F_A to simulate the distributed load. We can obtain the bending moment as $M_A = F_A \cdot L/3$. Figure 2.6b shows the free-body diagram with the beam cut at mid-span, where the bending moment is $M_B = F_B \cdot L/6$. Thus, we get the same conclusion.

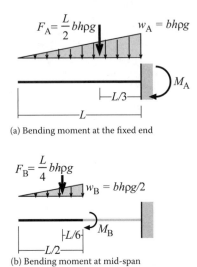

FIGURE 2.6 Cantilever beam with linearly varying depth.

3.1 The reactions at the supports are shown in Figure 3.1a. Thus, we obtain the shear force shown in Figure 3.1b, which implies that option 1 is correct. The bending moment diagram is shown in Figure 3.1c, implying that options 2 and 3 are correct. The deformation is shown in Figure 3.1d, which implies that option 5 is correct. As we have seen in Figure 3.1.13a, however, the deflection does not reach a maximum at point C. Option 4 is NOT correct.

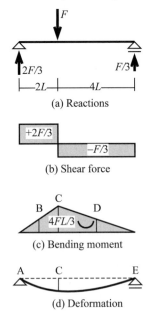

FIGURE 3.1 Simply supported beam.

Answers to Problems   329

3.2 Recalling $V = dM/dx$, we obtain the shear force for each bending moment diagram as shown in Figure 3.2. The corresponding external forces and reactions are shown in Figure 3.3. For the cases of questions (2) and (3), we should expect couples where the bending moment diagrams are discontinuous. The correct answers are 1c, 2d, 3e, 4a and 5b.

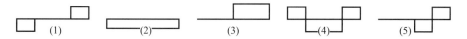

FIGURE 3.2 Shear force diagrams.

FIGURE 3.3 External forces and reactions.

3.3 As we have seen in Figure 3.2.7c, the maximum bending moment is

$$M = \frac{wL^2}{8} = \frac{12 \times 6^2}{8} = 54 \text{ kN-m} = 54 \times 10^6 \text{ N-mm}$$

As discussed in Section 2.5, the section modulus is

$$Z = \frac{bh^2}{6} = \frac{100 \times 180^2}{6} = 54 \times 10^4 \text{ mm}^3$$

The maximum stress in the beam is

$$\sigma = \frac{M}{Z} = \frac{54 \times 10^6}{54 \times 10^4} = 100 \text{ N/mm}^2$$

The correct answer is 2.

3.4 Imagine a beam of Figure 3.4 subjected to a uniformly distributed load w and a concentrated load R_B. The deflection of the beam at the center is

$$v = \frac{5wL^4}{384EI} - \frac{R_B L^3}{48EI}$$

Because the center of the beam in the problem is supported by a roller, v should be zero. This results in

$$R_B = \frac{5}{8} wL$$

On the other hand, the equilibrium of forces in the vertical direction requires

$$wL - 2R_A - R_B = 0$$

which gives

$$R_A = \frac{3}{16} wL$$

Therefore, we have $\frac{R_A}{R_B} = \frac{3}{10}$ and the correct answer is e.

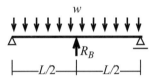

FIGURE 3.4 Beam continuous over two equal spans.

3.5 We shall call the reactions caused by the concentrated force R_B and R_C (Figure 3.5a). The equilibrium of moment around point C gives $R_B = 2F/3$ and the equilibrium of forces gives $R_C = F/3$ (Figure 3.5b). Cutting the beam at point A as shown in Figure 3.5c, we have $M_A = FL/3$. On the other hand, Figure 3.6a shows the replacement of the distributed force with a concentrated force wL. Figure 3.6b shows the reactions. Cutting the beam at point A as shown in Figure 3.6c, we have $M_A = wL^2/3$. The correct answer is $F = wL$. Option 3 is correct. For reference, the shear force and the bending moment diagrams are shown in Figures 3.5d,e and 3.6d,e.

3.6 Equilibrium of moments around point B or C gives the reactions as shown in Figure 3.7a. The shear force and bending moment diagrams are shown in Figures 3.7b and c. Note that these diagrams are equivalent to those of cantilever beams (Figure 3.7d). Because point C is supported by a pin (Figure 3.7e), we need to rotate it as shown in Figure 3.7f, which leads to a deflection at point A: $v = \frac{2FL^3}{3EI}$. Option 4 is correct.

3.7 Figure 3.8a shows the reactions and the deformed shape. Beam DC deforms as shown in Figure 3.8b, where the deflection of node C is $\frac{FL^3}{3EI}$. The forces and reactions acting on beam CBA are the same as those in problem 3.6. Therefore, the deflection of node A is obtained adding the contributions (Figure 3.8c).

$$v = \frac{FL^3}{3EI} + \frac{2FL^3}{3EI} = \frac{FL^3}{EI}$$

Answers to Problems

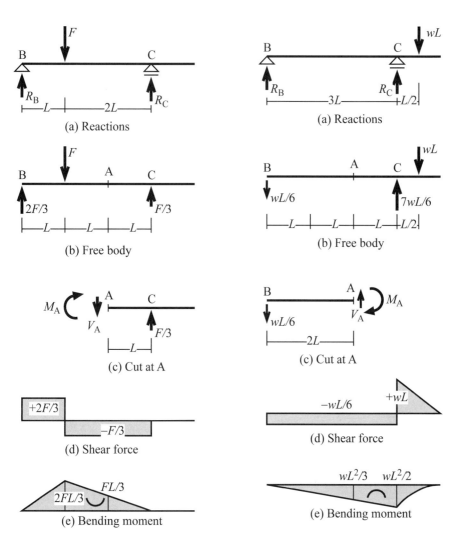

FIGURE 3.5 Contribution of concentric force.

FIGURE 3.6 Contribution of distributed force.

Option 3 is correct. (Summation of deflection is permitted if the slope of the beam is small enough. In Figure 3.8, deformations are exaggerated.)

3.8 Figures 3.9a, b, and c show the moment diagrams caused by one, two, and five couples. Note that the bending moment decreases as we increase the number of couples. The answer is, therefore, 'no moment, no deflection.' The uniformly distributed couples are equivalent to the horizontal forces uniformly distributed on the top and the bottom of the beam (Figure 3.9d). They cause shear deformation only.

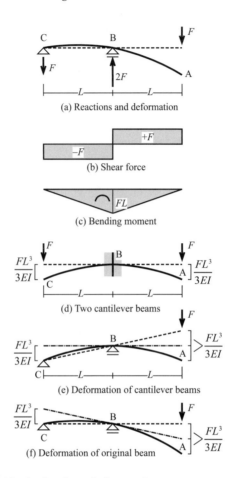

FIGURE 3.7 Beam with a load at the end of an overhang.

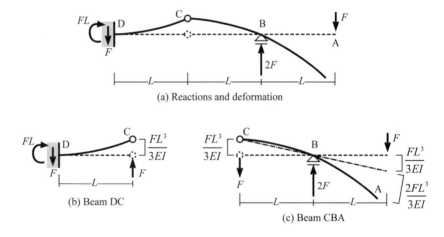

FIGURE 3.8 Beam with a hinge at mid-span and a concentrated load at the end of an overhang.

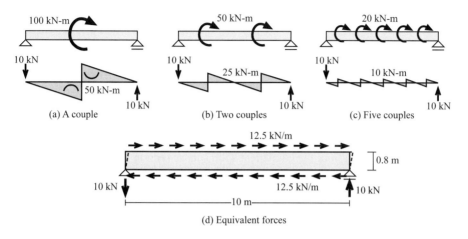

FIGURE 3.9 Uniformly distributed couples.

4.1 The moment of inertia of the section in Figure 4.5.1a is $I = a^4/12$, whereas that of Figure 4.5.1b is

$$I = \int y^2 dA = 2 \times \int_0^{a/\sqrt{2}} y^2 \times (\sqrt{2}a - 2y)dy$$

$$= 2 \times \left[\frac{\sqrt{2}ay^3}{3} - \frac{y^4}{2} \right]_0^{a/\sqrt{2}} = 2a^4 \times \left(\frac{1}{6} - \frac{1}{8} \right) = \frac{a^4}{12}$$

Therefore, the stiffness is the same. On the other hand, the section modulus of Figure 4.5.1a is $Z = a^3/6$, whereas that of Figure 4.5.1b is

$$Z = \frac{I}{a/\sqrt{2}} = \frac{\sqrt{2}a^3}{12} < \frac{a^3}{6}$$

The flexural strength ($M = Z\sigma_t$) of the section in Figure 4.5.1a is larger than that of the section in Figure 4.5.1b. In other words, the section in Figure 4.5.1a is stronger than that in Figure 4.5.1b.

4.2 The moment of inertia of the pipe is

$$I = \frac{\pi}{4}(500^4 - 480^4) \approx \frac{\pi}{4} \times 94.2 \times 10^8 \text{ mm}^4$$

Its cross-sectional area is

$$A = \pi(500^2 - 480^2) \approx \pi \times 19600 \text{ mm}^2$$

The radius of the equivalent solid circular section is

$$r = \sqrt{19600} = 140 \text{ mm}$$

Its moment of inertia is

$$I = \frac{\pi}{4} 140^4 \approx \frac{\pi}{4} \times 3.84 \times 10^8 \text{ mm}^4$$

Because $94.2/3.84 \approx 24.5$, Option 3 is correct.

4.3 The cross-sectional area of the I-section is

$$A = 300 \times 2 \times 24 + (700 - 2 \times 24) \times 13 = 22.9 \times 10^6 \text{ mm}^2$$

The side length of the equivalent solid square section is

$$a = \sqrt{22.9 \times 10^6} \approx 151 \text{ mm}$$

Its moment of inertia is

$$I = \frac{151^4}{12} \approx 4.36 \times 10^7 \text{ mm}^4$$

Because $194/4.36 \approx 44.5$, Option 5 is correct. The I-section is a very efficient shape for bending about an axis perpendicular to its web.

4.4 The shear stress caused by a shear force of $V = 1200$ kN at the center of the section is

$$\tau = \frac{3}{2} \times \frac{1200 \times 10^3}{400 \times 300} = 15 \text{ N/mm}^2$$

The shear stresses at the edges are zero. Option 2 is correct.

5.1 Couple M causes a uniform bending moment over the height of column AB as shown in Figure 5.1a. A horizontal force at node C causes the bending moment shown in Figure 5.1b. Horizontal and vertical forces at node

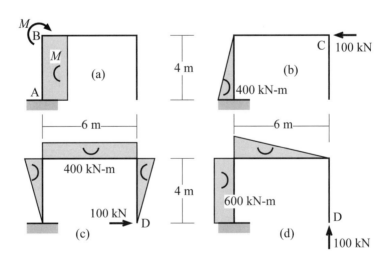

FIGURE 5.1 Bending moment diagrams.

D cause the bending moments shown in Figures 5.1c and d. Because the bending moment at A is to be zero, we obtain the equation

$$-M + 400 + 0 + 600 = 0 \quad \text{or} \quad M = 1000 \text{ kN}$$

Option 1 is correct.

5.2 The horizontal and vertical forces, F_x and F_y, cause the bending moments shown in Figures 5.2a and b. Because the bending moment at A should be zero, we get $F_x = F_y$. Option 2 is correct.

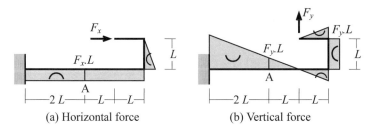

FIGURE 5.2 Bending moment diagrams.

5.3 Figure 5.3a shows the possible reactions. Because the horizontal reaction at support A is zero, the shear force (and, therefore, the bending moment) in column AB must be zero. Thus, we conclude that only option (a) is acceptable.

We can compute the reactions in Figure 5.3a as follows. First, force equilibrium in the horizontal direction leads to $R_{Dx} = F$. Second, moment equilibrium around support D leads to $R_{Ay} = (1/6)F$. Finally, force equilibrium in the vertical direction leads to $R_{Dy} = F - R_{Ay} = (5/6)F$. Thus, we obtain the bending moment diagram shown in Figure 5.3b. If A was a pin-support and D a roller-support as shown in Figure 5.3c, option (c) would be correct. In this case, the deflections of the portal are much larger. You should check it using GOYA-P and think the reason.

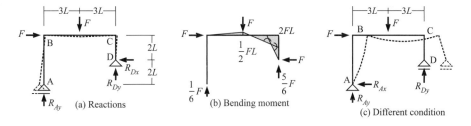

FIGURE 5.3 Portal frame.

5.4 Figure 5.4 shows shear force diagrams that correspond to the bending-moment diagrams in Figure 5.7.6. Figure 5.5 shows forces and reactions that correspond to the shear-force diagrams in Figure 5.4. Option (a) is incorrect because the frame is subjected to a horizontal force. Option (b) is incorrect because the magnitude of the vertical force is 8 kN. Options (c) and (d) seem to satisfy the boundary conditions shown in Figure 5.7.5. If we examine Figure 5.7.6d carefully, we notice that the bending moment at the left end of the beam is larger than that at mid-span. This is plausible if the flexural rigidity of the column is much larger than that of the beam, but not for this case. Option (c) is most plausible. (If you input very large Young's modulus for the columns in GOYA-P, you will obtain the moment diagram shown in Figure 5.7.6d.)

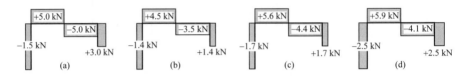

FIGURE 5.4 Shear force diagrams.

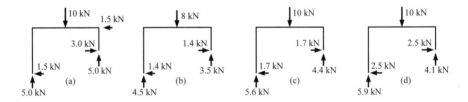

FIGURE 5.5 Forces and reactions.

5.5 The shear forces in columns AC and BD are

$$V_{AC} = V_{BD} = \frac{140+100}{4} = 60 \text{ kN}$$

as shown in Figure 5.6a. The horizontal force at the roof is, therefore, $60 \times 2 = 120$ kN as shown in Figure 5.6a. The shear forces in columns CE and DG are

$$V_{CE} = V_{DG} = \frac{150+170}{4} = 80 \text{ kN}$$

The total shear force in the first story is $80 \times 2 = 160$ kN. We conclude that the horizontal force at the second floor is

$$F_D = 160 - 120 = 40 \text{ kN}$$

and that option 5 is *incorrect*. Let us check the remaining options. Moment equilibrium at each node leads to the bending moment diagram of the beams shown in Figure 5.6b. Thus, we get

$$V_{AB} = \frac{2 \times 140}{8} = 35 \text{ kN}, \quad V_{CD} = \frac{2 \times 250}{8} = 62.5 \text{ kN} \quad \text{and} \quad V_{EG} = \frac{2 \times 170}{8} = 42.5 \text{ kN}$$

and conclude that options 1 and 2 are correct. As we did in Example 5.5.1, we determine the axial force in column DG

$$P_{DG} = V_{AB} + V_{CD} = 35 + 62.5 = 97.5 \text{ kN}$$

Option 3 is correct.
Finally, the reaction at support G is

$$R_{Gy} = V_{AB} + V_{CD} + V_{EG} = 35 + 62.5 + 42.5 = 140 \text{ kN}$$

Option 4 is correct.

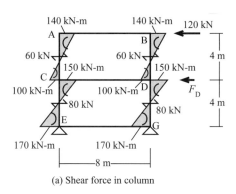
(a) Shear force in column

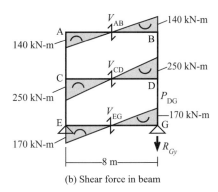
(b) Shear force in beam

FIGURE 5.6 Two-story frame.

5.6 If we recall Eq. 5.5.5, we get

$$\delta_2 = \frac{L^3}{24EI} \times 2F \quad \text{and} \quad \delta_1 = \frac{L^3}{24 \times 2EI} \times (2F + F)$$

Rearranging,

$$\frac{\delta_1}{\delta_2} = \frac{3}{4}$$

Option 5 is correct.

5.7 Figure 5.7a shows the probable reactions. Moment equilibrium around support A leads to

$$8F \times 2L + 4F \times L - R_{Ey} \times 4L = 0 \quad \text{or} \quad R_{Ey} = 5F$$

Figure 5.7b shows the free-body diagram for the right part of the frame. Moment equilibrium around node C leads to

$$R_{Ex} \times 2L - 5F \times 3L = 0 \quad \text{or} \quad R_{Ex} = 7.5F$$

We conclude that the bending moment at node D is

$$M_D = R_{Ex} \times 2L = 15FL$$

Option 3 is correct. Figure 5.7c shows the bending moment diagram.

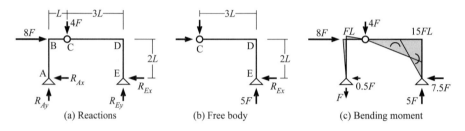

FIGURE 5.7 Three-hinged frame.

6.1 Column CH needs to resist the snow on a tributary area of $20 \times 20 = 400$ ft². The axial force of column CH is

$$P = (400 \text{ ft}^2) \times (1{,}000 \text{ lbf/ft}^2) = 4.0 \times 10^5 \text{ lbf}$$

The moment of inertia of the column section is

$$I = \frac{\pi}{4}\left(R_{out}^4 - R_{in}^4\right) = \frac{\pi}{4}(4.0^4 - 3.0^4) = 137 \text{ in}^4$$

The column will buckle as shown in Figure 6.1a. The buckling force is given by Equation 6.1.22.

$$P_{cr} = \frac{\pi^2 EI}{L^2} = \frac{3.14^2 \times (30{,}000 \times 10^3) \times 137}{(40 \times 12^2)} = 1.76 \times 10^5 \text{ lbf} < 4.0 \times 10^5 \text{ lbf}$$

We conclude that this design is inadequate.

6.2 The column will buckle as shown in Figure 6.1b. The buckling force is given by Equation 6.1.24 (twice the previous answer).

$$P_{cr} = \frac{2\pi^2 EI}{L^2} = 2 \times 1.76 \times 10^5 = 3.52 \times 10^5 \text{ lbf} < 4.0 \times 10^5 \text{ lbf}$$

We conclude that this design is still inadequate.

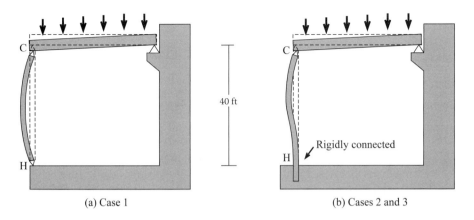

(a) Case 1   (b) Cases 2 and 3

FIGURE 6.1 Buckling failure.

6.3 Because the columns are connected by a rigid beam, we expect that all the three columns will buckle at the same time. Noting that the columns carry snow from a total tributary area of $20 \times 40 = 800$ ft² (the broken line in Figure 6.2), the total snow load is

$$P = (800 \text{ ft}^2) \times (1{,}000 \text{ lbf/ft}^2) = 8.0 \times 10^5 \text{ lbf}$$

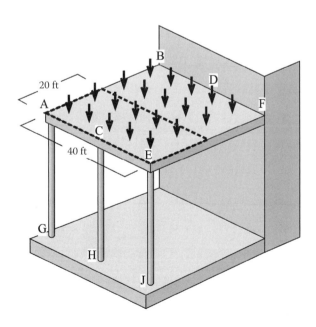

FIGURE 6.2 Tributary area for case 3.

The buckling force is three times the previous answer.

$$P_{cr} = 3 \times \frac{2\pi^2 EI}{L^2} = 3 \times 3.52 \times 10^5 = 10.6 \times 10^5 \text{ lbf} > 8.0 \times 10^5 \text{ lbf}$$

Now, the design appears to be adequate. But we need to think further. The actual columns may not be perfectly vertical. The connections between the columns and the foundation beams may not be perfectly rigid. Therefore, it would be wise to use larger pipes so that P_{cr} would be larger than 1.5 times P. Snow load can be quite demanding and we hear of failures that occur because an engineer has underestimated the load or overestimated the strength.

List of Symbols

a	length of a segment (mm or in.)
A	area of cross section (mm^2 or $in.^2$)
b	width of cross section (mm or in.)
C	couple (N-mm or lbf-in.) or compressive force (N or lbf)
e	deformation (Section 1.2) or eccentricity (Section 6.1) (mm or in.)
E	Young's modulus (N/mm^2 or $lbf\text{-}in.^2$)
F	force (N or lbf)
h	depth of cross section (mm or in.)
I	second moment of a section or moment of inertia (mm^4 or $in.^4$)
L	length (mm or in.)
M	bending moment (N-mm or lbf-in.)
P	axial force (N or lbf)
R	reaction (N or lbf)
S	first moment of a section (mm^3 or $in.^3$)
T	tensile force (N or lbf)
u	displacement (mm or in.)
v	deflection (mm or in.)
V	shear force (N or lbf)
w	load per unit length (N/mm or lbf/in.)
y_0	distance from the bottom of a section to the neutral axis (mm or in.)
Z	section modulus (mm^3 or $in.^3$)
α	rotation of a spring (rad)
Δ	deflection (mm or in.)
ε	unit strain (no unit)
ϕ	unit curvature (/mm or /in.)
θ	angle or slope (rad)
σ	unit stress (N/mm^2 or $lbf\text{-}in.^2$)

USER'S MANUAL FOR GOYA

How to Get GOYA

Access one of the following Web sites to download GOYA:
    http://kitten.ace.nitech.ac.jp/ichilab/mech/en/
    ftp://ftp.ecn.purdue.edu/sozen/GOYA

How to Start GOYA-T

You can use most types of computers (Windows, Macintosh, or Linux). Double click the icon "GOYA-T.jar." If you do not get the window shown in the following intext figure, visit http://www.java.com/ and install JAVA.

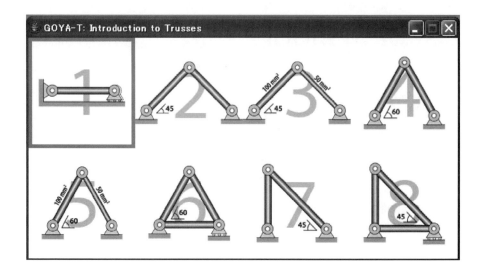

How to Use GOYA-T1

Click the icon at the upper left of the window shown in the preceding intext figure to get the following window.

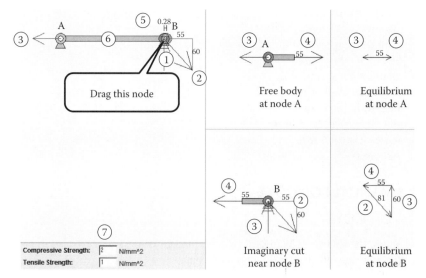

① Drag node B as you want. (Not the red arrow.)

② The red arrow shows the external force. The red numbers show the x- and y- components of the force.

③ The green arrows show the reactions.

④ The blue arrow shows the axial force.

⑤ The black number shows the displacement of node B.

⑥ The cross-sectional area of the bar is assumed to be 200 mm². The color changes to blue if its tensile strength and to red if its compressive strength is exceeded.

⑦ Type the desired strength of the material.

List of Symbols

How to Use GOYA-A

Double click the icon "GOYA-A.jar" to find the window shown in the following intext figure.

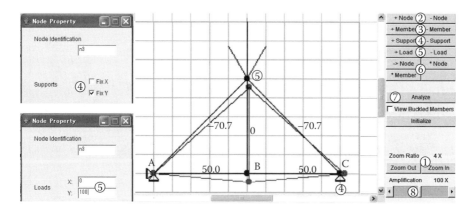

① Push the 'Zoom In' button to have a close-up view of the canvas.

② To add a node, push the '+Node' button and click on the canvas.
   To remove a node, push the '−Node' button and click the node.

③ To add a member, push the '+Member' button and drag between the nodes, e.g., AB and BC.
   Do not drag between A and C directly.
   To remove a member, push the '−Member' button and drag between the nodes.

④ To add a roller support, push the '+Support' button, check 'Fix Y' in a new window, and click the node. To remove a support, push the '−Support' button and click the node.

⑤ To add a load, push the '+Load' button, type values in a new window, and click the node.
   To remove a load, push the '−Load' button and click the node.

⑥ To move a node, push the '−>Node' button and drag the node.
   To display the properties of a node, push the '*Node' button and click the node.
   To change the properties of a member, push the '*Member' button and drag between the nodes.

⑦ Push the 'Analyze' button to have the result.

⑧ To amplify or reduce the deformation, slide the 'Amplification' bar.

How to Use GOYA-C (GOYA-S for Simple Beam is Similar to GOYA-C)

Double click the icon "GOYA-C.jar" to find the window shown in the following intext figure.

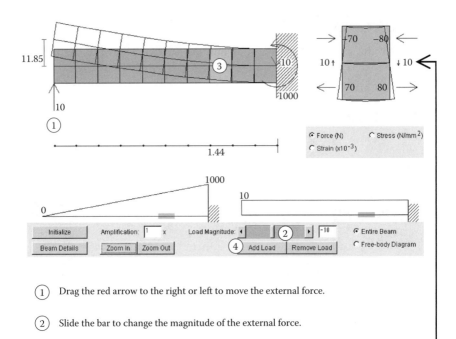

① Drag the red arrow to the right or left to move the external force.

② Slide the bar to change the magnitude of the external force.

③ Click a segment to see the forces acting on the segment here.

④ Click the 'Add Load' button to create another load.

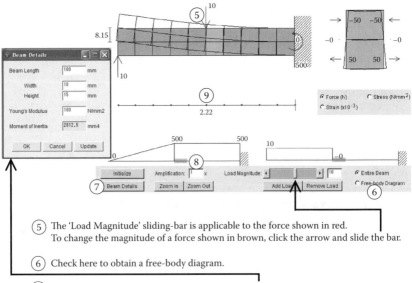

⑤ The 'Load Magnitude' sliding-bar is applicable to the force shown in red. To change the magnitude of a force shown in brown, click the arrow and slide the bar.

⑥ Check here to obtain a free-body diagram.

⑦ Click the 'Beam Detail' button and type in a new window to change the beam properties.

⑧ Type any number and hit the 'Enter' key to amplify or reduce the deformation.

⑨ This number shows the deflection at the selected segment.

List of Symbols

How to Use GOYA-I

Double click the icon "GOYA-I.jar" to find the window shown in the following intext figure.

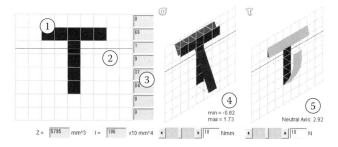

① The black squares constitute the section. To change a black square to white or vice versa, click the square.

② The red line shows the neutral axis.

③ These numbers show the contribution of each row to the moment of inertia.

④ The numbers after 'min=' and 'max=' show the maximum compressive and tensile stresses, respectively.

⑤ This number shows the shear stress at the neutral axis.

⑥ If you want to make another window, double click the 'GOYA-I.jar' icon, again.

How to Use GOYA-N

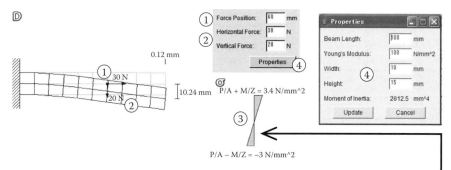

① To move the location of the force, drag the red dot or type a value.

② To change the magnitude of the force, drag the tip of the red arrow or type a value.

③ This figure shows the stress distribution in the section at the fixed end.

④ To change the beam length etc, push the 'Properties' button, type the values in a new window, and push the 'Update' button.

How to Use GOYA-F2 (GOYA-F1 and F3 are Similar to GOYA-F2)

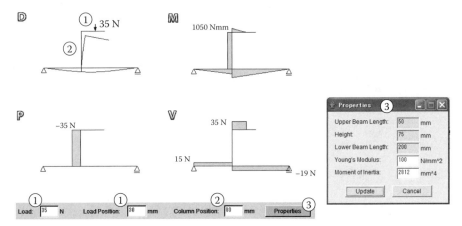

① To change the location and magnitude of the force, drag the red arrow or type values.

② To move the location of the column, drag the column or type a value.

③ To change the beam length etc, push the 'Properties' button, type the values in a new window, and push the 'Update' button.

How to Use GOYA-F4

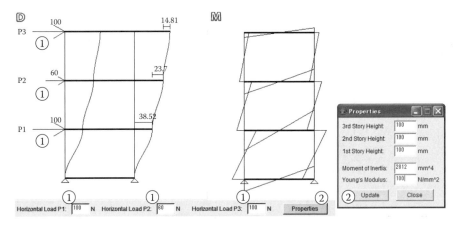

① To change the magnitude of the force, drag the red arrow or type values.

② To change the story height, push the 'Properties' button, type the values in a new window, and push the 'Update' button. 'Moment of Inertia' and 'Young's Modulus' apply to all the columns.

Index

A

Archimedes, 43, 206
Axial force diagrams, portal frame, 251, 253, 255, 257, 259
Axial forces
  angle of 30°, 20
  angle of 45°, 19
  angle of 90°, 18
  bar in tension/compression, 5
  basic concepts, 14–15
  beam and truss response similarities, 183, 186–187
  buckling, 291, 293–294
  couple, 121–122
  defined, 5
  first moment, 201
  frames, 227, 229–230, 232, 234, 236
  multistory frame, 276
  portal frame, 251
  second moment, 206
  shear force, 95, 97
  simple bent frame, 238–240, 242–247
  statically determinate structures, 61–67
  statically indeterminate frames, 261–262
  strength of symmetrical trusses, 21
  stresses in a beam, moment effect on, 126
  symmetrical trusses, 18–21, 26
  three-hinged frames, 280
  trusses, hands-on design, 54–55, 58
  trusses with three members, 40–41
  two elements at various orientations, trusses, 26–29
  unsymmetrical trusses, 31, 33–35, 38
Axis
  buckling, 301–302
  continuously deformable model, 143
  first moment, 199, 201
  strains, 130–131
  stresses, 130

B

Baby elephants, 24, *see also* Design your own
Bar in tension/compression
  axial forces, 14–15
  examples, 10–14
  GOYA-T, 3, 8–10
  trusses, 3–15
Beams, *see specific type*
Beauty, multistory frame, 272
Bell, E. T., 206
Bending and shear stresses, *see also* Shear stress; Stresses
  center of gravity computation, 206
  centroid of a section, determination, 203–206
  first moment, 199–206
  flanges, 209–210
  problems, 225–226, 333–334
  second moment and section modulus, 206–215
  section resisting moment, optimum proportions, 215
  shear stress, 218–224
  Styrofoam beam, construction and test, 216–218
  web, 209–210
Bending moment
  beam and truss response similarities, 187
  buckling, 296
  concentrated load, effects, 167, 170
  continuously deformable model, 147, 149–150, 302–303, 307
  couple, 118, 120–121
  defined, 79
  and deformation, 88–89
  design your own, 86–87
  distributed loads, 108
  exercises, 85–86
  experiment, 85
  frames, 228–229, 236
  GOYA-C, 79, 80, 83, 86, 87
  moments and deflections, cantilever beams, 79–89
  multistory frame, 273–274, 278
  portal frame, 251–252, 261
  second moment, 207, 213
  several concentrated load, effects, 173–174, 178–179
  shear force, 89–93, 95, 97
  shear stress, 220
  simple bent frame, 238, 247
  spring models, 139–140, 142

statically indeterminate frames, 261–262, 264–265, 268–270
strains in a beam, moment effect on, 132–134
stresses in a beam, 123–130
Bending moment diagrams
continuously deformable model, 306
couple, 120
distributed loads, 107, 109–111, 113, 115–117
frames, 229–230, 232
multistory frame, 273–275
portal frame, 252–253, 255, 257, 259
several concentrated load, effects, 171, 176
shear force, 99–103
simple bent frame, 242–250
statically indeterminate frames, 266–270
three-hinged frames, 280, 282
Bernoulli studies, 130
Bone, human, 126–128
Bridges, 35, 213
Buckling
continuously deformable models, 302–308
problems, 309, 338–340
simple models, 291–302
trusses, hands-on design, 52–58

C

Cantilever beams
and bending beams, 130
bending moment, 85
buckling, 296
continuously deformable model, 145
several concentrated load, effects, 171
simple bent frame, 237–238, 240, 243–244
spring models, 138
statically indeterminate frames, 266
timber beam, construction and test, 191, 192
Cantilever beams, moments and deflections
bending moment, 79–89
cantilever beams and bending beams, 130
continuously deformable model, deformation, 143–155
couple, 118–123
deformation, 88–89, 135–155
design your own, 86–87, 103–104, 117–118, 129, 153–154
distributed load, 106–118
effects on strains, 130–135
effects on stresses in a beam, 123–130
internal, 104–105
mini-game, 153
problems, 156–159, 322–328
review, 121–122
shear force, 89–105

spring model, deformation, 135–142
unit curvature, 154–155
Cantilever columns
buckling, 291–294, 301–302
continuously deformable model, 306
frames, 234
Cathédrale Saint-Pierre de Beauvais, 1
Center of gravity computation, 206
Centroids
beam and truss response similarities, 183
of a section, determination, 203–206
several concentrated load, effects, 178
stresses in a beam, moment effect on, 126
Chord force, 183–185
Coffee break
axial forces in a truss, 25
cantilever and bending beams, 130
center of gravity computation, 206
Collapse, 36, 52, *see also* Failure
Columns, *see* Cantilever beams
Compression/tension, bar in
axial forces, 14–15
examples, 10–14
GOYA-T, 3, 8–10
trusses, 3–15
Concentrated loads
couple, 118
examples, 164–166
exercises, 167, 170
GOYA-S, 161, 166, 167
moments and deflections, simply supported beams, 161–170
point of zero slope detection, 169
Concrete, 126–128
Constants of integration, 303–304
Contest, wood and string, 69, 72–73
Continuously deformable models
design your own, 153–154
examples, 145, 147–148, 149–153
exercises, 147–148, 151, 153
GOYA-C, 148, 151, 153
mini-game, 153
moments and deflections, cantilever beams, 143–155
spring models, 136
unit curvature, 154–155
Continuously deformable models, buckling
examples, 305–308
exercises, 304–305, 308
fundamentals, 302–304
GOYA-U4, 304–305
GOYA-U5, 307
Conversations
axial forces, 14–15
bending moment and deformation, 88–89

Index

calculating reactions, 48–51
internal shear force, 104–105
unit curvature, 154–155
Coulomb studies, 130
Couple
  bending moment, 83
  continuously deformable model, 147
  example, 120–121
  GOYA-C, 118, 120
  moments and deflections, cantilever beams, 118–123
  review, 121–122
  sensing, 193
  several concentrated load, effects, 172
Crew and de Salvio studies, 130
Critical load, buckling, 299–300
Crow-bar action, 92–93
Curvature, *see also* Unit curvature
  beam and truss response similarity, 186–187
  concentrated loads, 165, 169
  continuously deformable model, 302–304
  defined, 145
  first moment, 199, 201
  second moment, 207–208
  several concentrated loads, 174

D

da Vinci, Leonardo, 25
Deflections, 228–229, *see also* Moments and deflections
Deformation
  bending moment, 84, 88–89
  design your own, 153–154
  examples, 138–142, 145, 147–148, 149–153
  exercises, 137, 138, 140, 147–148, 151, 153
  GOYA-C, 137–141, 148, 151, 153
  mini-game, 153
  moments and deflections, cantilever beams, 135–155
  multistory frame, 278
  portal frame, 252, 257–259
  shear force, 92
  unit curvature, 154–155
Degrees of freedom, 69–71
de Salvio, Crew and, studies, 130
Design of trusses, hands-on approach, 51–58
Design your own
  bending moment, 86–87
  continuously deformable model, 153–154
  deformation of beam, 153–154
  distributed load, 117–118
  effects on stresses, 129
  frames, 236
  several concentrated loads, effects, 180–181

shear force, 103–104
symmetrical trusses, 25, 30–31
three member trusses, 48
unsymmetrical trusses, 39
Desk example, 218–219
Diagonals, 181
Displacements, 11
Distributed loads
  couple, 120
  design your own, 117–118
  examples, 108–117
  GOYA-C, 106, 107, 111–114, 117–118
  moments and deflections, cantilever beams, 106–118
Domes, 35, *see also* Nagoya Dome

E

Earthquakes, 213
Eccentricity, 293–294
Effects on strains
  examples, 134–135
  exercises, 135
  GOYA-C, 133, 134–135
  moments and deflections, cantilever beams, 130–135
Effects on stresses
  design your own, 129
  examples, 126–129
  GOYA-C, 123, 125, 129
  moments and deflections, cantilever beams, 123–130
Eiffel Tower, 1
Elastic buckling, 294
Elephants, *see* Design your own
Elongation
  concentrated load, effects, 166
  frames, 228
  statically determinate structures, 62, 67
  symmetrical trusses, two elements at various orientations, 25
  thinner and longer bars, 10
  trusses with three members, 41
  unsymmetrical trusses, 36
Equations of deflected shape of beams, 190, 191
Equilibrium
  angle at 30°, 20
  angle at 45°, 19
  bending moment, 84
  distributed loads, 107
  first moment, 201
  frames, 234
  fundamentals, 2
  multistory frame, 275
  portal frame, 257

shear force, 91
shear stress, 220, 222–223
simple bent frame, 237
stable trusses, 70
statically determinate structures, 61, 63–66
symmetrical trusses, 19–20, 25–26
three-hinged frames, 280, 282–283
trusses, hands-on design, 54
trusses with three members, 40–41, 43–45
two elements at various orientations, 25
unsymmetrical trusses, 34, 37–38
Equilibrium diagram, 31
Equilibrium of moments
  concentrated load, effects, 161, 167
  several concentrated load, effects, 171
  shear force, 90–91
  shear stress, 221
Euler solutions, 130, 291, 301–302, 304
Examples
  analysis of statically determinate type, 62–67
  bar in tension/compression, 10–14
  concentrated load, effects, 164–166
  continuously deformable model, 145, 147–148, 149–153
  continuously deformable models, buckling, 305–308
  couple, 120–121
  deformation of beam, 138–142, 145, 147–148, 149–153
  distributed load, 108–117
  effects on strains, 134–135
  effects on stresses, 126–129
  first moment, 202–203
  frames, 229–236
  second moment and section modulus, 210–215
  several concentrated loads, effects, 174–179
  shear force, 100–103
  shear stress, 223–224
  simple models, buckling, 298–302
  spring model, 138–142
  symmetrical trusses, 19–24, 27–30
  three member trusses, 45–47
  unsymmetrical trusses, 33–34, 37–38
Exercises
  analysis of statically determinate type, 65
  bending moment, 85–86
  concentrated load, effects, 167, 170
  continuously deformable models, 147–148, 151, 153, 304–305, 308
  deformation of beam, 137, 138, 140, 147–148, 151, 153
  effects on strains, 135
  response similarities, beams and trusses, 188
  several concentrated loads, effects, 171
  shear force, 99

simple models, buckling, 294, 295, 298
spring model, 137, 138, 140
three member trusses, 45
Experiments
  bending moment, 85
  centroid of a section, determination, 203–206
  first moment, 203–206
External forces
  bending moment, 79
  couple, 121–122
  symmetrical trusses, two elements at various orientations, 26–27
  trusses, hands-on design, 54–55

F

Failure, 22–23, 28, *see also* Collapse
Finite region, 145
First moment
  bending and shear stresses, 199–206
  centroid of a section, determination, 203–206
  examples, 202–203
  experiments, 203–206
  GOYA-I, 199–200, 203
First-order derivative of the deflection, 143
Flanges, 209–210
Forces, *see also specific type*
  concentrated load, effects, 164
  continuously deformable model, 148–149, 307
  first moment, 201
  moment and vector relationship, 50–51
  several concentrated load, effects, 171, 176
  shear force, 89
  spring models, 137
  stresses in a beam, moment effect on, 125–126
  three-hinged frames, 282–284
  unit measurement, 6–7
Frames
  concepts, 227–236
  design goals, 271–272
  design your own, 236, 250
  illustrative design examples, 272–275
  joint, 241–242
  multistory frame, 271–279
  portal frame, 251–261
  problems, 286–289, 334–338
  simple bent, 236–250
  statically indeterminate frame, 261–271
  three-hinged frames, 279–286
Free body, 16, 90
Free-body diagrams, *see also* Imaginary cut
  axial forces, 12
  bending moment, 79, 83
  buckling, 296, 301–302
  concentrated load, effects, 161, 163–164

Index 351

continuously deformable model, 302
defined, 16
portal frame, 251, 253, 257
shear stress, 221
simple bent frame, 243–244
statically determinate structures, 66
symmetrical trusses, two elements at various orientations, 26
three-hinged frames, 280, 282–284
trusses with three members, 40–41, 43

G

Galileo, Galilei, 130
Gordon, J.E., 9
GOYA (general), 341
GOYA-A
  analysis of statically determinate type, 58–61, 65, 67
  building trusses, 72
  response similarities, beams and trusses, 185, 187
  using, 343
GOYA-C
  bending moment, 79, 80, 83, 86, 87
  continuously deformable beam model, 148, 151, 153
  couple, 118, 120
  deformation of beams, 137–141, 148, 151, 153
  distributed load, 106, 107, 111–114, 117–118
  frames, 227
  several concentrated loads, effects, 171
  shear force, 89, 99, 104
  spring beam model, 137–141
  strains, effects on, 133, 134–135
  stresses, effects on, 123, 125, 129
  using, 343–344
GOYA-F4, 346
GOYA-F2 (F1/F3 similarly), 346
GOYA-I
  first moment, 199–200, 203
  second moment and section modulus, 207, 209, 215
  using, 345
GOYA-N
  frames, 227–228
  using, 345
GOYA-S
  concentrated load, effects, 161, 166, 167
  response similarities, beams and trusses, 185
  several concentrated loads, effects, 171, 172, 176, 180, 181
  using, 343–344

GOYA-T
  bar in tension/compression, 3, 8–10
  symmetrical trusses, 16, 17, 21, 23, 25, 26–27
  three member trusses, 39, 45, 47
  two elements at various orientations, 25, 26–27
  two elements of equal size at 90°, 16, 17, 21, 23
  unsymmetrical trusses, two elements, 36–39
GOYA-T1, 342
GOYA-U1, 294
GOYA-U2, 295
GOYA-U3, 298
GOYA-U4, 304–305
GOYA-U5, 307

H

Heat, 35–36, 279
Horizontal forces
  buckling, 302
  frames, 227, 230, 232
  multistory frame, 274–275, 277
  portal frame, 251, 253–255, 257, 259
  simple bent frame, 239
  statically indeterminate frames, 262, 264–265
  symmetrical truss, two elements of equal size at 90°, 18
  three-hinged frames, 283
  unsymmetrical trusses, 33–34
Human bone, 126–128

I

Imaginary cut, see also Free-body diagrams
  axial forces, 12
  bar in tension/compression, 5
  defined, 5
  statically determinate structures, 66
Imperial system (measurement), 7
Infinitesimal region
  continuously deformable model, 144
  first moment, 201–203
  second moment, 206, 212
  shear stress, 219, 221
Interactions, see Conversations
Interested readers
  joint, 241–242
  moment, force, and vector relationship, 50–51
  point of zero slope detection, 169–170
  stress-strain relationship, 10
Internal forces
  couple, 121–122
  statically indeterminate frames, 261–262
  symmetrical trusses, two elements at various orientations, 26–27
Internal shear force, 104–105

J

Joints, *see also* Nodes
  defined, 16
  method of, 65
  multistory frame, 273

L

Lateral forces, 254, *see also* Horizontal forces
Lever principle, 43–44

M

Maximum moment, 168
Maxwell, James Clerk, 190
Mega-Pascal (MPa), measure, 6
Method of joints, 65
Method of sections, 65, 67–68
Mini-elephants, *see* Design your own
Mini-games, 153, 180
Möbius, A. F., 69
Moment
  anticlockwise, 44
  beam and truss response similarities, 183
  clockwise, 44
  defined, 43–44, 79, 82
  force and vector relationship, 50–51
  trusses with three members, 43–44
Moment diagrams
  concentrated load, effects, 167
  shear force, 97
  simple bent frame, 237, 240
  three-hinged frames, 283, 285
Moment equilibrium
  bending moment, 84
  buckling, 294
  distributed loads, 107
  frames, 234
  several concentrated load, effects, 179
  three-hinged frames, 280, 282
Moment of inertia
  beam and truss response similarities, 187
  buckling, 291, 299
  concentrated load, effects, 166
  continuously deformable model, 305
  second moment, 207, 211–215
  shear stress, 223–224
  strains in a beam, moment effect on, 133, 135
  timber beam, construction and test, 193
Moments and deflections, cantilever beams
  bending moment, 79–89
  cantilever beams and bending beams, 130
  continuously deformable model, deformation, 143–155

couple, 118–123
  deformation, 88–89, 135–155
  design your own, 86–87, 103–104, 117–118, 129, 153–154
  distributed load, 106–118
  effects on strains, 130–135
  effects on stresses in a beam, 123–130
  internal, 104–105
  mini-game, 153
  problems, 156–159, 322–328
  review, 121–122
  shear force, 89–105
  spring model, deformation, 135–142
  unit curvature, 154–155
Moments and deflections, simply supported beams
  build your own, 180–181
  cantilever beams, 191, 192
  concentrated load, effects, 161–170
  couple, sensing, 193
  equations of deflected shape of beams, 190, 191
  measuring Young's modulus, wood, 189–190
  mini-game, 180
  moment of inertia, effects of, 193
  point of zero slope detection, 169
  problems, 194–197, 328–333
  reciprocity theorem, 190–191, 192
  response similarities, beams and trusses, 181–188
  simply supported beams, 190
  timber beam, construction and test, 188–193
Multistory frame
  design goals, 271–272
  examples, 275–279
  exercises, 279
  GOYA-M, 279
  illustrative design examples, 272–275

N

Nagoya Dome, 1, 2
Navier studies, 130
Nervous mini-elephants, *see* Design your own
Neutral axis
  centroid of a section, 205
  first moment, 199, 201–203
  second moment, 206–207, 212
  shear stress, 221–222, 224
  stresses in a beam, moment effect on, 130
Newton, Isaac, 206, 307
Newton's third law, 16
Nodes, *see also* Joints
  defined, 16
  multistory frame, 273–275
  unsymmetrical trusses, 31, 37–38

Index

P

Palladio, Andrea, 1
Parabola, 107, 222
Pine, 126–128
Pin joint, 16
Pin supports
   bar in tension/compression, 3
   defined, 3
   horizontal reaction, 11
   multistory frame, 273
   portal frame, 259
   statically determinate structures, 59, 61
   statically indeterminate frames, 268
   trusses with three members, 39
Point of zero slope detection, 169–170
Portal frame
   examples, 259–261
   exercises, 261
   fundamentals, 251–259
   GOYA-P, 251, 253, 254, 257, 259
Principle of the lever, 43–44
Problems
   answers to all, 311–340
   bending and shear stresses, 225–226, 333–334
   buckling, 309, 338–340
   frames, 286–289, 334–338
   moments and deflections, cantilever beams, 156–159, 322–328
   moments and deflections, simply supported beams, 194–197, 328–333
   trusses, 73–78, 311–322

R

Reactions/reaction forces
   calculating, 48–51
   concentrated load, effects, 167
   defined, 4
   frames, 234
   multistory frame, 277
   portal frame, 253, 261
   several concentrated load, effects, 171, 173
   statically determinate structures, 59, 64
   symmetrical truss, two elements of equal size at 90°, 17, 19
   trusses with three members, 43, 46–47
Readers, *see* Interested readers
Reciprocity theorem, 190–191, 192
Response similarities, beams and trusses
   exercises, 188
   GOYA-A, 185, 187
   GOYA-S, 185
   moments and deflections, simply supported beams, 181–188

Review, couple, 121–122
Rigid-discrete models, 302
Roller supports
   bar in tension/compression, 5
   defined, 3
   horizontal reaction, 11
   portal frame, 259
   simple bent frame, 249
   statically determinate structures, 59, 61
   statically indeterminate frames, 268
   three-dimensional trusses, 71
   trusses with three members, 39, 41

S

Safe domain
   symmetrical trusses, 22–24, 27
   trusses with three members, 45–46
   two elements at various orientations, 27, 29–30
   unsymmetrical trusses, 33, 37–38
Safety
   buckling, 299–300
   multistory frame, 271
Seasonal stresses
   three-hinged frames, 279
   unsymmetrical trusses, 35–36
Second moment and section modulus
   bending and shear stresses, 206–215
   examples, 210–215
   flanges, 209–210
   GOYA-I, 207, 209, 215
   section resisting moment, optimum proportions, 215
   web, 209–210
Second-order differential of the deflection, 144
Section modulus
   continuously deformable model, 307
   second moment, 209–214
   stresses in a beam, moment effect on, 125, 127, 129
Section resisting moment, optimum proportions, 215
Sections, method of, 65, 67–68
Serviceability, 272
Several concentrated loads, effects
   build your own, 180–181
   examples, 174–179
   exercises, 171
   GOYA-C, 171
   GOYA-S, 171, 172, 176, 180, 181
   mini-game, 180
   moments and deflections, simply supported beams, 171–181

Shallow trusses, 57, *see also* Trusses
Shear deformation/distortion, 187–188
Shear-force diagrams
   concentrated load, effects, 167
   couple, 118, 120
   distributed loads, 106–107, 109–111, 113, 115–117
   portal frame, 251, 253, 255, 257, 259
   simple bent frame, 248
   three-hinged frames, 280
Shear forces
   beam and truss response similarities, 183
   concentrated load, effects, 167
   continuously deformable model, 152
   couple, 121
   defined, 89
   design your own, 103–104
   distributed loads, 108
   examples, 100–103
   exercises, 99
   frames, 229–230, 232
   GOYA-C, 89, 99, 104
   internal, 104–105
   moments and deflections, cantilever beams, 89–105
   multistory frame, 275, 277–278
   several concentrated load, effects, 173, 178–179
   simple bent frame, 237, 242–247
   statically indeterminate frames, 261–262
Shear stress, *see also* Bending and shear stresses; Stresses
   bending and shear stresses, 218–224
   examples, 223–224
Simple bent frames
   design your own, 250
   examples, 242–248, 249–250
   exercises, 248
   fundamentals, 236–241
   GOYA-D, 248
   GOYA-L, 236, 250
   joint, 241–242
Simple models, buckling
   examples, 298–302
   exercises, 294, 295, 298
   fundamentals, 291–294
   GOYA-U1, 294
   GOYA-U2, 295
   GOYA-U3, 298
Simply supported beams, 190
SI system (measurement), 6
Skyscrapers, 298
Slopes
   concentrated load, effects, 165, 169–170
   continuously deformable model, 150

couple, 121
statically indeterminate frames, 264
Smoother diagrams, 106
Space station truss, *3*
Spaghetti strand, 305–308
Spring model, deformation
   examples, 138–142
   exercises, 137, 138, 140
   GOYA-C, 137–141
   moments and deflections, cantilever beams, 135–142
Squat trusses, 55–56, *see also* Trusses
Stable trusses, 68–72
Statically determinate structures, 34–35
Statically determinate trusses
   analysis of, 58–68
   examples, 62–67
   exercises, 65
   GOYA-A, 58, 65, 67
Statically indeterminate frame
   examples, 266–270
   exercises, 270–271
   fundamentals, 261–265
   GOYA-P, 265, 267, 270
Steel, 126–128
Stiffness
   buckling, 293–294, 298, 300
   compared to strength, 9
   multistory frame, 274
Storm effects, 213
Story drift, 278
Strain-hardening limit, 10
Strains
   bar in tension/compression, 6
   concept of, 7–8
   defined, 6
   examples, 134–135
   exercises, 135
   GOYA-C, 133, 134–135
   moments and deflections, cantilever beams, 130–135
   small, 8
   stress relationship, 10
   unit measurement, lack of, 7
Strength
   buckling, 294
   compared to stiffness, 9
   continuously deformable model, 307
Stresses, *see also* Bending and shear stresses; Shear stress
   bar in tension/compression, 6
   concept of, 7
   defined, 6
   design your own, 129
   effects of moments, in a beam, 123–130

examples, 126–129
GOYA-C, 123, 125, 129
moments and deflections, cantilever beams, 123–130
seasonal, 35–36
second moment, 214–215
strain relationship, 10
unit of measurement, 6–7
Styrofoam beam, construction and test, 216–218
Symbols, list of, 341
Symmetrical trusses, two elements at various orientations
  concepts, 25–31
  design your own, 30–31
  examples, 27–30
  GOYA-T, 25, 26–27
Symmetrical trusses, two elements of equal size at 90°
  concepts, 16–25
  design your own, 25
  examples, 19–24
  GOYA-T, 16, 17, 21, 23

T

Talking, *see* Conversations
Tall trusses, 55–56, *see also* Trusses
Tensile forces, 221–223
Tensile stresses, 307–308
Tension/compression, bar in
  axial forces, 14–15
  examples, 10–14
  GOYA-T, 3, 8–10
  trusses, 3–15
Ten springs model, 136–137
Three-dimensional trusses, 71
Three-hinged frames
  examples, 283–285
  exercises, 285–286
  fundamentals, 279–283
  GOYA-H, 280, 282, 283, 285
Three member trusses
  calculating reactions, 48–51
  concepts, 39–51
  design your own, 48
  examples, 45–47
  exercises, 45
  GOYA-T, 39, 45, 47
  vectors defined, 50–51
Timber beam, construction and test
  cantilever beams, 191, 192
  couple, sensing, 193
  equations of deflected shape of beams, 190, 191
  measuring Young's modulus, wood, 189–190
  moment of inertia, effects of, 193
  moments and deflections, simply supported beams, 188–193
  reciprocity theorem, 190–191, 192
  simply supported beams, 190
Timoshenko studies, 130
Trusses
  analysis of statically determinate type, 58–68
  axial forces, 14–15, 25
  bar in tension/compression, 3–15
  building, 72–73
  calculating reactions, 48–51
  defined, 1–3
  design, hands-on approach, 51–58
  design your own, 25, 30–31, 39, 48
  Nagoya Dome, *2*
  problems, 73–78, 311–322
  stable type, 68–72
  symmetrical, 16–31
  three members, 39–51
  two elements at various orientations, 25–31
  two elements of equal size at 90°, 16–25
  unsymmetrical, two elements, 31, 33–39
  vectors defined, 50–51
  wood and string, contest using, 69, 72–73
Two spring model
  buckling, 296–298, 300–301
  deformation of a model, 138–139, 142–143

U

Unit curvature, *see also* Curvature
  basic concepts, 154–155
  continuously deformable model, 144–146, 148–150, 152
  strains in a beam, moment effect on, 131–135
Unstable trusses, 68–69, *see also* Stable trusses
Unsymmetrical trusses, 34
Unsymmetrical trusses, two elements
  concepts, 31–39
  design your own, 39
  examples, 33–34, 37–38
  exercises, 36–37
  GOYA-T, 36–37

V

Vector, 50–51
Vertical forces
  distributed loads, 107
  frames, 227, 230, 232
  multistory frame, 274, 277
  portal frame, 253, 255, 257, 261
  simple bent frame, 239

statically indeterminate frames, 265
three-hinged frames, 283, 285
Vitruvius Pollio, Marcus, 271
von Leibnitz, Gottfried W., 206

W

Web, 209–210
Web members, 181, 183, 184
Welds, bad, 36
Wood
  beam, shear force, 97
  properties, Young's modulus effect, 53
  and string, contest using, 69, 72–73

Y

Young, Thomas, 9
Young's modulus
  beam and truss response similarities, 186
  buckling, 291
  concept of, 9
  first moment, 201
  statically indeterminate frames, 262
  strains in a beam, moment effect on, 131, 133
  Styrofoam beam, 217
  trusses, hands-on design, 53, 57–58
  unsymmetrical trusses, 35